Ocean Thermal Energy Conversion (OTEC) - Past, Present, and Progress

Edited by Albert S. Kim and Hyeon-Ju Kim

Published in London, United Kingdom

IntechOpen

Supporting open minds since 2005

Ocean Thermal Energy Conversion (OTEC) - Past, Present, and Progress
http://dx.doi.org/10.5772/intechopen.86591
Edited by Albert S. Kim and Hyeon-Ju Kim

Contributors
Weimin Liu, Yun Chen, Yunzheng Ge, Lei Liu, Alejandro García Huante, Yandy Rodríguez Cueto, Erika Paola Garduño Ruíz, Ricardo Efraín Hernández Contreras, Alonso Pérez Pérez, Mauricio Andres Latapi Agudelo, Graciela Rivera Camacho, Estela Cerezo-Acevedo, Victor Manuel Romero Medina, Miguel Alatorre Mendieta, Jessica Guadalupe Tobal Cupul, Elda Gomez Barragan, Juan Francisco Bárcenas Graniel, Enrique Celestino Carrera Chan, María Fernanda Sabido Tun, Hyeon-Ju Kim, Michael Petterson, Ho-Saeng Lee, Seung-Taek Kim, Jung-In Yoon, Abu Bakar Jaafar, Mohd Khairi Abu Husain, Azrin Ariffin

First published in London, United Kingdom, 2020 by IntechOpen
IntechOpen is the global imprint of INTECHOPEN LIMITED, registered in England and Wales, registration number: 11086078, 7th floor, 10 Lower Thames Street, London, EC3R 6AF, United Kingdom
Printed in Croatia

British Library Cataloguing-in-Publication Data
A catalogue record for this book is available from the British Library

Additional hard and PDF copies can be obtained from orders@intechopen.com

Ocean Thermal Energy Conversion (OTEC) - Past, Present, and Progress
Edited by Albert S. Kim and Hyeon-Ju Kim
p. cm.
Print ISBN 978-1-78985-571-5
Online ISBN 978-1-78985-572-2
eBook (PDF) ISBN 978-1-83880-521-0

An electronic version of this book is freely available, thanks to the support of libraries working with Knowledge Unlatched. KU is a collaborative initiative designed to make high quality books Open Access for the public good. More information about the initiative and links to the Open Access version can be found at www.knowledgeunlatched.org

For EU product safety concerns:
IN TECH d.o.o., Prolaz Marije Krucifikse Kozulić 3, 51000 Rijeka, Croatia, info@intechopen.com or visit our website at intechopen.com.

Meet the editors

Dr. Albert S. Kim is currently a full professor at the Department of Civil and Environmental Engineering at the University of Hawaii at Manoa. He has been teaching in the department since 2001, after he earned his MS (1997) and PhD (2000) in Civil and Environmental Engineering from the University of California at Los Angeles. Dr. Kim's scientific accomplishments include the National Science Foundation Faculty Early Career (CAREER) Award in 2005, the University of Hawaii Regents' Medal for Excellence in Research in 2006, and the University of Hawaii Regents' Medal for Excellence in Teaching in 2017. Professor Kim has published more than 60 peer-reviewed journal papers and six book chapters, and edited two open access books.

Hyeon-Ju Kim has been the leader of R&D and Industrialization of Seawater Resource and its Utilization Plant at the Korea Research Institute of Ships and Ocean Engineering (KRISO) since 2005, and is Senior Director of the Offshore Plant and Marine Energy Research Division. He has managed DOWA (Deep Ocean Water Application) and OTEC (Ocean Thermal Energy Conversion) projects funded by the Korean government since 2000. His team carried out a demonstration test by floating a 1 MW OTEC plant in 2019. He holds a doctoral degree in Ocean Engineering from Pukyoung National University, and is an ocean engineering specialist licensed in Korea. His professional experience ranges from theoretical analysis to experimental evaluation of sea water utilization systems for food, energy, and water.

Contents

Preface

The 21st century does not have a readily available non-fossil energy source that is large enough to be exploited on the requisite scale. The seven types of renewable energy include solar, wind, hydroelectric, geothermal, ocean, hydrogen, and biomass processes. Among them, the ocean can produce two kinds of energy: thermal energy using the depth-induced temperature difference between surface and deep seawater, and kinetic energy using the waves, ebbs, and flows of the tides. The former is specifically called ocean thermal energy conversion (OTEC), and comes from an original idea written in a science fiction novel *20,000 Leagues Under the Sea* (1870) by a French novelist, poet, and playwright, Jules Gabriel Verne.

Those scientists who have devoted their passion to OTEC include Jacques Arsene d'Arsonval and his former student Georges Claude for open-cycle OTEC from the 1880s to the 1930s, and J. H. Anderson and J. H. Anderson, Jr. for closed-cycle OTEC in the 1960s. In the United States, the Natural Energy Laboratory of Hawaii Authority (NELHA) was established in the 1970s and became a leading OTEC test facility. The current status of OTEC has a number of obstacles to overcome, such as technological breakthroughs of performance versus footprint, uncertainty in cost analysis and finances, and sociocultural/political issues. Because an OTEC plant is an extensive, expensive infrastructure to build for the benefit of everyone, the geographical adaptation of the technology is much more important than global standardization.

Now, in 2020, is the right time to reinvestigate past experiences, present activities, and future perspectives of OTEC in various holistic aspects—from the fundamental, microscopic science level to long-term operations/maintenance in harmony with human engineering and Mother Nature. This book shares state-of-the-art OTEC technology, especially from the 7th Ocean Thermal Energy Conversion (OTEC) Symposium (26–27 September, 2019, Busan, Korea).

Albert S. Kim
Professor,
Department of Civil and Environmental Engineering,
The University Of Hawaii At Manoa, Honolulu,
Hawaii, USA

Dr. Hyeon-Ju Kim
Senior Director of Offshore Plant and Marine Energy Research Division,
Korea Research Institute of Ships and Ocean Engineering,
Yuseong-gu, Daejon, Korea

Section 1

Perspective and Overview

Chapter 1

Can Ocean Thermal Energy Conversion and Seawater Utilisation Assist Small Island Developing States? A Case Study of Kiribati, Pacific Islands Region

Michael G. Petterson and Hyeon Ju Kim

Abstract

The deployment of a land-based Ocean Thermal Energy Conversion (OTEC) plant in South Tarawa, Kiribati, Pacific Islands Region, in 2020/2021, represents a major technical achievement, alongside an international development opportunity. Pacific Small Island Developing States (PSIDS) are archipelago nations with small land areas and large oceanic exclusive economic zones. Geographical isolation and large transport distances make economic development a challenge. A lack of affordable and reliable energy in many PSIDS is a development inhibitor. PSIDS are situated within the areas of highest ocean thermal potential in the world. Temperature differences between surface and 1 km depth waters, are in excess of 24°C. Regional geology and tectonics allow access to deeper, colder, waters within few kilometres of many shorelines, and close to market. Seawater Utilization technologies can catalyse varied industrial development (e.g., fresh water/aquaculture/agriculture/mineral salts). The KRISO (Korean Research Institute of Ships and Ocean Engineering)-Government of Kiribati OTEC partnership is already 7 years old (2013–2020) and has involved extensive negotiations, awareness raising programmes, and inclusive collaboration. The project will test OTEC technologies and explore a range development opportunities for Kiribati. The programme could become a role model for the application of the concept of 'Interconnected Geoscience'.

Keywords: ocean thermal energy conversion, OTEC, international development, Kiribati, green energy, Pacific

1. Introduction: why OTEC, seawater utilisation, and SIDS?

This paper examines aspects of the application of ocean thermal energy conversion (OTEC) and seawater utilisation within a Pacific Small Island Developing States (or PSIDS) development context. OTEC was first proposed in 1881 by a French physicist, Jacques-Arsene d'Arsonval, and the first OTEC plant was built in Cuba, in 1930, by Georges Claude. The principles of OTEC are discussed in later sections. It is worthwhile, at this stage, considering questions such as "why has such an old technology taken so long to be realised on a large scale?" and "why deploy at this time within a Pacific Islands context?"

There are a number of potential replies to the first question. Many technological ideas and inventions do not end up as large-scale commercial successes. There may be long incubation periods for some technologies before their application need becomes apparent, or the technology may not allow development on a large or mass-produced scale until scientific advances occur. The idea of space travel or mobile communication devices, for example, was common in science fiction, long before they were technologically realised. With respect to energy, there has been, and remains, an abundance of hydrocarbon energy, with oil in particular, being highly transportable and flexible as an energy source. The advent of climate change and global warming social and political movements, particularly since the 2015 COP 21 meeting in Paris, France [1], are heralding the gradual demise of fossil fuels and the rise of less polluting renewable energies. This change in thinking, policy, and economics has allowed OTEC to become, again, a renewable energy source that may, finally, come of age. Technical and commercialisation challenges remain for OTEC, particularly in the sphere of large (100 MW plus) fully ocean-deployed energy platform development, and this will impede progress for some time to come. Only small (<1 MW) land-based, ocean-adjacent, OTEC systems have been developed thus far, as experimental plants or provision of small-scale energy, drinking water, agriculture/aquaculture, or space heating/cooling units in places such as France, Hawaii, India, Mexico, and South Korea. There remains a wide gap between commercialisation need (for large electricity generation plants) and current OTEC technical capabilities.

The second question may, at first, appear cryptic, but does, on analysis, make a degree of logical sense. Why, from all the world's markets would an advanced country such as South Korea choose a small Pacific atoll island nation to be the target of, potentially, the world's first-ever 1 MW OTEC plant? Why not China, the USA, Canada, South America, or the European Union? One answer is scale. Large developed countries, or even medium-sized emerging countries, require far more electricity than a small OTEC plant can provide. Then there is geography. An OTEC plant requires oceanic temperature conditions that are only met year-round, in tropical and subtropical waters. So SIDS and Pacific SIDS (PSIDS), from the viewpoint of this paper, start to become appealing. Many PSIDS are surrounded by enormous ocean energy potentials (if only the energy can be tapped) and geological/topographic conditions that allow for rapid access to deeper, cold water, alongside the warmest ocean surface temperatures in the world. PSIDS in particular have underdeveloped electricity generation and supply infrastructure, much of which is old, expensive, inefficient, unreliable, and dependent on imported oil. Total electricity demand for the smaller PSIDS is low, between 5 and 20 MW. Therefore, the development of even a 1 MW OTEC plant within a small PSIDS can add significant amounts of energy to the grid, reduce reliance on imported oil, generate new skills and employment opportunities, and have additional benefits in the area of drinking water provision, refrigeration/air conditioning, agriculture, aquaculture, and, even, mineral salt/cosmetic manufacture. In theory, there are many development 'wins' for the deployment of OTEC within a small PSIDS. Alongside the concept of OTEC is the concept of seawater utilisation, which describes the manifold applications of seawater such as in the fields of aquaculture, agriculture, and mineral salt and cosmetic manufacture. Deep seawater has a number of characteristics that make it useful, such as a lack of potentially harmful pollutants and organic substances and a chemical composition that promotes aspects of human health.

This paper will examine a number of aspects of OTEC deployment within the Pacific Islands region, particularly focusing upon the 1-year period deployment and testing of an OTEC plant in South Tarawa, Kiribati. It will critically examine the application of new science and technology to the Pacific region from the philosophical lens of 'interconnected geoscience' [2] and the sustainable development goals (SDGs) [3] and develop the conversation of developmental needs and futures of PSIDS and where OTEC could fit in.

2. OTEC, seawater utilisation, and the sustainable development goals

Seawater utilisation plants use seawater as a base resource to produce food, energy, and drinkable water through ocean thermal energy conversion systems such as seawater cultivation, seawater energy, and seawater desalination technologies. Seawater energy and seawater utilisation plants can be developed in tropical SIDS to utilise its seawater as a heat source to produce renewable energy and heat, water, and food. These technologies can assist with the sustainable development of coastal communities.

The Korean Research Institute of Ships and Ocean Engineering (KRISO) has led research and development on OTEC and seawater utilisation of discharged deep seawater since 2010. A 20 kW OTEC pilot plant was designed and fabricated as a prototype model of the 1 MW demonstration OTEC plant (to be deployed in Kiribati in 2020). Results and discoveries made from the prototype OTEC/seawater utilisation plant have been used to design the 1 MW OTEC Kiribati plant.

The application of discharged deep seawater from a land-based OTEC plant, or from individual cold water pipes, alongside technologies for desalination, has been developed by KRISO and the Korean R&D team. Seawater desalination plants with carefully designed features can enhance, and control, the constituent seawater mineral balance/composition to make it particularly useful for public health, cosmetics, mineral salt manufacture, and other industrial applications. Seawater utilisation plants are green technologies, reducing CO_2 emissions, and supplying renewable energy. They can be used to develop a 'blue infrastructure' (technologies based on the utilisation of the neighbouring ocean) in coastal regions, which help to promote the UN sustainable development goals (see [4–8], for further details).

If the 1-year duration experiment in Kiribati can evolve into a long-term OTEC + seawater utilisation plant, it has the potential to address a wide series of the sustainable development goals [3]. In particular the SDGs 7 (affordable and clean energy), 6 (clean water and sanitation), 9 (industrial innovation and infrastructure), and 13 (climate action) will be addressed through the provision of renewable and affordable, reliable energy, and the development of innovative technologies and related industries, particularly from a PSIDS perspective. If the OTEC/seawater utilisation plants work inclusively with local people, offering education and training to enable the localization of technologies with time, and if industries such as agriculture, aquaculture, mineral salt, and cosmetic manufacture, and so on, are developed, then SDGs such as 1 (no poverty), 2 (zero hunger), 3 (good health and wellbeing), 4 (quality education), 8 (decent work and economic growth), and 11 (sustainable cities and communities) can be promoted. These are serious claims and will not occur without a great deal of long-term investment and effort (which is not guaranteed at the time of publication). However, there is a genuine vision alongside the mere technical deployment of a 1 MW OTEC plant. If locally based agriculture/aquaculture industries develop, they can address the limited diet available to I-Kiribati people, promoting sustainable food resources and healthy eating. The provision of high-quality education and training alongside locally owned ancillary industries to OTEC will address the areas of poverty, quality livelihoods, and education opportunities. The growth of new, green technology-based industry can contribute towards decent work, economic growth, and sustainable cities/communities. Of course, none of this will occur without longer-term investment and planning.

3. An 'interconnected geoscience' approach to OTEC in the Pacific Islands region

The approach taken to the introduction of OTEC into the Pacific Islands region is a good model for the application of sustainable development principles and the concept of *interconnected geoscience* (**Figure 1**, [2]). Sustainable development was defined in detail in the 1989 Brundtland Report, Brundtland [9], which coined the *phrase* 'meeting the needs of today without compromising the needs of tomorrow' and demonstrated the dynamic links between society, economy, environment and politics/governance.

Interconnected geoscience is a conceptual model of geoscience/technological/engineering application to international development. A definition of interconnected geoscience is 'a philosophy that combines geoscience expertise with an equivalent expertise/consciousness in the understanding of developmental situations, conditions, and context, including the integration of diverse world views/wisdom and values, placing development-goals at the heart of the interconnected-approach' [2].

International development requires a complex series of human, knowledge, and often technical interactions and activities undertaken for the purpose of improving the quality of life of the world's less empowered and least-wealthy. It involves aspects of nation building such as economic strengthening, infrastructure development, job creation, improved social welfare such as health and education, and improved governance. It is impossible to reduce this grand aspiration to only simple reductionist activities, such as the building of a bridge, road, or railway or even the installation of an OTEC plant *alone*. Of course reductionist activities can and have been undertaken alone, almost as an isolated, totally independent project. They may

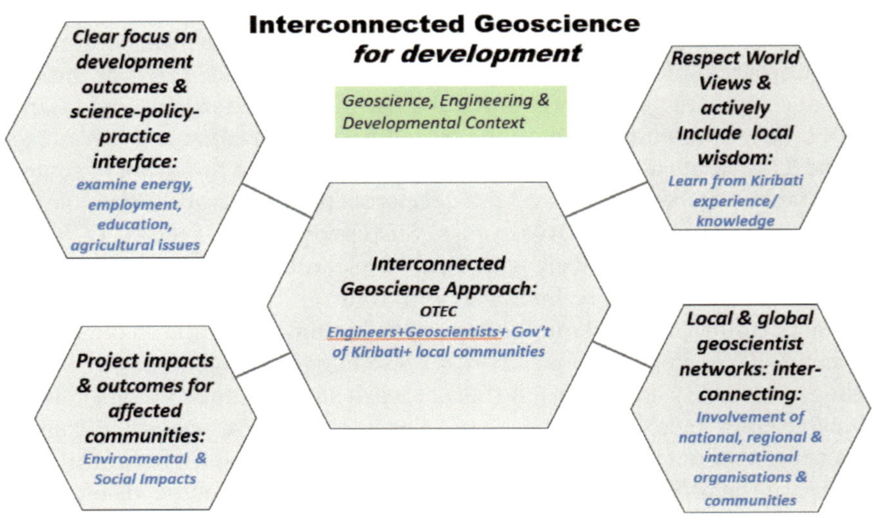

Figure 1.
Interconnected geoscience is a concept advocating the application of excellent geoscience/engineering/technical work to international development that includes contextual conditions such as community, level of development, and local world views/wisdom. This diagram summarises the key 'interconnected' components of the Kiribati OTEC-seawater utilisation programme (adapted from Petterson [2]).

indeed create good results, such as improved communications. However, if activities and interventions are not planned, taking the context of a development situation into account, difficulties can arise, and situations may result where more harm than good is effected, e.g. through increased corruption, money wastage, local communities adversely affected, and so on.

Work by authors such as Gill and Bullough [10], Stewart and Gill [11], and Stewart [12] points towards a deeper, non-reductionist, holistic, and inclusive guiding philosophy, which encourages the application of geoscience (or OTEC technology), within a development framework that takes account of the local situational context in terms of culture, community, society, governance, planning, environment, and local wisdom/world views.

The deployment of the OTEC plant within Kiribati is the pinnacle of a series of actions that have occurred from late 2013 to 2020. The process has included a wide range of stakeholders including the Korean government, KRISO, the Fiji-based Korean Pacific diplomatic mission, KOICA (Korean Aid), the Pacific Community (SPC), and representatives from Pacific Island governments, particularly interactions with the governments of Kiribati, Tuvalu, and the Marshall Islands, community representatives, private sector, and others. Preparatory activities have included numerous one-to-one and one-to-many meetings, workshops, training sessions, scientific surveys, inter-government and community negotiations, and training/education courses at institutes such as the Kiribati Marine Training School. These activities and stakeholder interactions have allowed for the gradual raising of awareness of all parties to many aspects of OTEC and its application potential within the Pacific Islands region.

One interesting aspect of this interaction is the connections between two quite distinctive and different cultures: the South Korean Asian culture and the Pacific Islander culture. Without descending into stereotypes, but simplifying the description of two cultures, this article highlights aspects of the respective cultures of Korea and the Pacific Islands. South Korea, prior to recent developments in China, is, perhaps, the country that has become 'developed' (in modernistic industrial economics terms) the fastest: from a post-conflict agrarian society following the 1950s Korean War to an industrial giant of today (the eleventh, out of 193, richest economy in 2019, [13]). South Korea prizes social values such as a strong work ethic, competition, high educational achievement, innovation, and industrial/commercial progress. Pacific Island cultural values are quite different to mainstream South Korean values. Pacific Islanders come from a largely agrarian, fishing, and hunter-gatherer society, which has become more urbanised during the twentieth and twenty-first centuries, whilst retaining a high element of a traditional lifestyle that extends back hundreds and thousands of years. The role of the community, tribe, and extended family is strong. Individualistic values and lifestyles, whilst existing, are not necessarily admired. Pacific societies can have strong societal hierarchies, and the role of a Chief, Paramount Chief, or equivalent, is of great importance. Most Pacific Islanders are committed to a Christian Church which also acts as a communal societal institution. Land ownership concepts are quite different to Asian or Western mindsets. Land is not individually but communally 'owned' (including reefs and oceans). The 'ownership' concept is better described as custodianship. The land is kept within communities and is an intrinsic part of communal identity and religion (particularly pre-Christian religious-philosophical beliefs which remain extant). The land is nurtured and tended and must be passed on to future generations in better health than the present time. Livelihoods and values, of course, do not remain static and, inevitably, change. The onset of the money economy and the nation state in the Pacific region has changed lifestyles and values, but longstanding beliefs and cultures remain strong. Many Pacific Islanders now live in modern urban centres such as Suva, Port Vila, and Apia and work in jobs similar to anywhere in a globalised world. However, many also retain traditional lifestyles and are strongly attached to a Pacific world view (see [14] or further details).

The OTEC project team, working with the Korean Diplomatic and KOICA missions in Fiji, has worked closely with the Pacific Community (SPC). SPC is a regional scientific and technical organisation, operative since 1947, with 22 Pacific Island countries and territories as members. The Pacific Community organisation is experienced in designing and delivering international development projects across the region. Development partnerships and a generous lead-in time (some 6 years) from concept to OTEC deployment in Kiribati, together with numerous initiatives from the Korean side, closely involving Pacific Islanders, have allowed the key tenets of *interconnected geoscience* to be deployed [2]. Specifically these interconnected characteristics include (1) the realisation of the OTEC technology and the identification of the optimal OTEC site locations from geoscientific and engineering principles, data generation, data modelling, and scenario development; (2) early project scoping discussions between the Korean side and SPC; (3) the delivery of c. 20 workshops in numerous locations involving Pacific Islanders, international agencies, and Korean stakeholders, with the objectives of introducing Korean thought and technology to a Pacific audience and providing platforms for Pacific Islanders, and others, to share knowledge and world views/wisdom; (4) working closely with Pacific Island governments who expressed a particular desire to become more heavily involved in OTEC issues (e.g. Tuvalu, the Marshall Islands, Kiribati); (5) working with the Kiribati government on all aspects of OTEC deployment in South Tarawa (Kiribati), particularly in relation to economic opportunities, land ownership, community awareness, governance, and environmental and social impacts; (6) designing the OTEC technology deployment and testing programme with Kiribati and Pacific needs in mind; and (7) increasing awareness of the developmental context of the Pacific region. Development, in a situation such as the OTEC deployment in Kiribati must be mindful of power asymmetries with respect to economic standing, relative levels of education, finance, and technology. The power asymmetry is quite acute when 'developed' countries work with PSIDS: in this case South Korea (11/193 in 2019 economic league tables) is working with Kiribati (191/193 in 2019 economic league tables) [13].

Project deployment will commence in 2020/21. There remains much work and learning to be undertaken. Projects and aspirations do not, of course, always run to plan. However, the preparation of the KRISO-OTEC programme with respect to the principles of *interconnected geoscience* has been a positive model.

4. Pacific Islands geography and geology

The Pacific Islands region occupies an area of close to 30 million km^2, mainly in the western and central Pacific, east and north of Australia, and north of New Zealand (**Figure 2**). Most island groups are in the southern hemisphere, with countries such as Kiribati, the Marshall Islands, Federated States of Micronesia, and Palau mainly in the northern hemisphere, close to the equator. Cook Islands, French Polynesia, and Pitcairn are the easternmost island groups (see [14] for further details). There are 22 PSIDS, some of which are independent nations such as Fiji, and some of which are territories of western countries, particularly France (e.g. New Caledonia). Papua New Guinea (PNG) is unusual in that it contains a large area of continent in addition to islands and has a relatively large land surface area (460,000 km^2) and population (>8 million people). If PNG is excepted from the

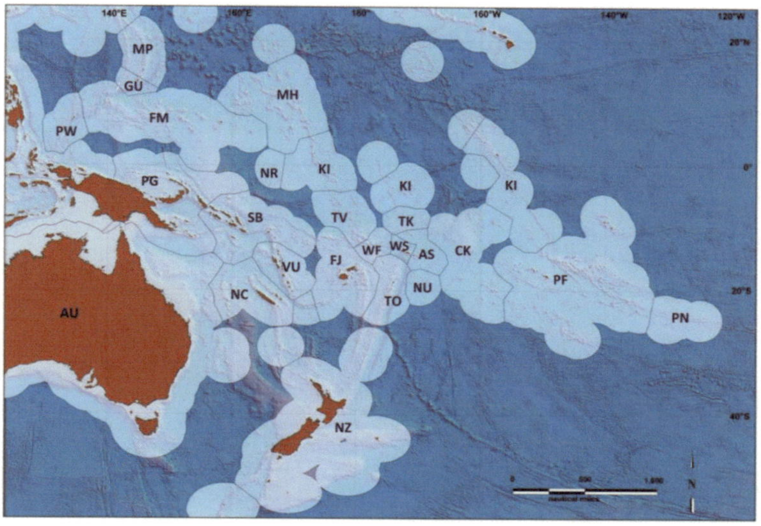

Figure 2.
Geography of the Pacific Islands region. Note the archipelago nature of most PSIDS with islands scattered over large areas of ocean. AS, American Samoa; AU, Australia; CI, Cook Islands; FM, Federated States of Micronesia; FJ, Fiji; PF, French Polynesia; GU, Guam; KI, Kiribati; MH, Marshall Islands; NR, Nauru; NC, New Caledonia; NU, Niue; NZ, New Zealand; MP, Marianas Islands; PG, PNG; PN, Pitcairn; PW, Palau; WS, Samoa; SB, Solomon Islands; TK, Tokelau; TO, Tonga; TV, Tuvalu; VU, Vanuatu; WF, Wallis and Futuna.

analysis, the remaining countries have a combined land surface area of c. 87.000km^2. The Pacific Islands geographical context therefore is similar to a country with a land area of Austria set in an ocean the size of Africa. Most PSIDS are archipelago nations, some with hundreds or over one thousand islands and tens of island groups within their jurisdiction. Many PSIDS are not only small island states but also large ocean states with exclusive economic zones extending up to 5 million km^2. Populations vary between 900,000 (Fiji) and 10,000 (Tuvalu).

The Pacific Islands region contains a high degree of geodiversity (**Figure 3**). The region is one of the best examples of a spectrum of oceanic tectonic phenomena in the world, displaying geological features such as ocean trenches, island arcs (islands formed from the subduction of one ocean plate beneath another), ocean plateaux (formed from rapid, voluminous mantle plume head eruptions), ocean basins (main component of oceans resulting from seafloor spreading), seamounts and seamount chains (extinct volcanoes produced as ocean plates pass over static mantle plumes), and rifted, submerged, and aseismic continental materials. The main plates are the enormous Pacific and Australian plates, together with a number of smaller 'microplates', such as the deep ocean basins south of PNG-Solomon Islands in **Figure 3**.

Arc-linear archipelago chains of islands characterise island arcs such as the Solomon Islands, Vanuatu, and Tonga (**Figure 3**). These islands are located on the subduction side of equally long curvilinear ocean trenches, up to c. 10–11 kms beneath sea level. The trenches mark the sites of interaction between the Australian and Pacific plates and produce earthquakes to depths of c. 700 kms and magnitudes of up to >8 on the Richter scale. Island arcs form the largest Pacific Islands which contain the larger human populations. The region contains two of the world's largest ocean plateaux: the Ontong Java and Manihiki Plateau (**Figure 3**). Much of the SW Pacific region comprises the deep ocean abyssal plains, some 4–6 kms deep. Volcanism is evident throughout the region and exhibits a range of styles from explosive to quiet (effusive). The region contains abundant seabed minerals which may, one day, form the basis of a significant industry in the region.

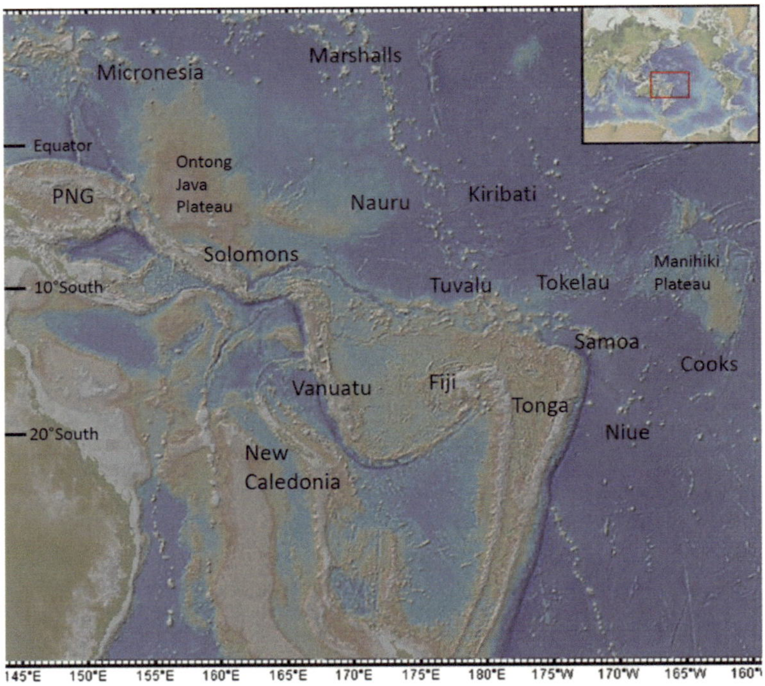

Figure 3.
The Pacific regions contain a wide geodiversity and present a range of ocean geological features, see text for details. Colour code: Deep blue/purple represents deepest ocean depths, pale blue represents shallower ocean, and browns and white/pale grey (within the ocean) represent topographic highs, the highest points of which form islands (acknowledgements to Google earth for the base topographic map).

Figure 4 illustrates an example of a typical atoll island, such as South Tarawa, Kiribati or the Marshall Islands. Islands tend to be up to tens of kilometres long by tens of metres to 1–2 kms wide. The islands form from the aggradation and erosion of coral reefs which develop around slowly subsiding, extinct volcanoes. Seamount slopes are very steep: often depths of 1–3 km beneath sea level are attained within 1–4 kms of the outer shoreline of atolls. Atoll islands themselves are composed of broken up coral in the form of rock blocks and boulders, but mainly sand and

Figure 4.
Photograph of a remote island atoll from the Gilbert Group of Kiribati, Pacific Islands region. Atoll islands form annular rings of low-lying (<1–4 m above sea level) made of sand, gravel and deeper igneous rock. An inner, shallower lagoon is separated from the deep ocean by the atoll islands (Photograph: Petterson).

gravel. The inner lagoon is shallow (tens of metres deep) and contains shifting sands and muds which are transported via tidal currents within the lagoon. Life on atolls is dependent on a freshwater lens which forms from rainfall: freshwater lies on top of seawater (as it is less dense) and forms drinkable groundwater. Groundwater is supplemented by rainwater harvesting. Atoll soils have a low fertility as they are largely composed of sand, mud, and gravel with limited organic material. They may be moderately saline. Coconut and breadfruit trees, alongside slow-growing swamp taro, are the main terrestrial edible atoll crops.

From an OTEC perspective, the geology and tectonics of the Pacific region are important in terms of determining the optimal location of OTEC plants. Firstly a market is needed, and these will be the larger towns of the Pacific region. Secondly, access to deep ocean, close to the market, is needed: Many sites fulfil this requirement with seamounts being outstanding examples. Thirdly, access to shipping and ports will be needed to maintain the OTEC plant. Natural disaster considerations are important as earthquakes, landslides, ocean-based landslides, volcanic eruptions, and extreme weather systems such as cyclones, can destroy an OTEC plant. Atoll island nations such as Tuvalu, Kiribati, and the Marshall Islands are situated upon seamounts (extinct volcanoes), with a relatively low earthquake risk, and quiet equatorial waters from where cyclones originate, but then away to the north or south, causing little/no damage to equatorial islands. Ocean climates tend to be calmer close to the equator.

5. Pacific Islands development needs

Readers are referred to Petterson and Tawake [14] for a more detailed analysis of this subject and to SPC [15], UN [16], and UNDP [17]. Traditionally the economy of the Pacific has relied on agriculture, fisheries, and traditional handicrafts/cottage industry. Countries such as PNG and Fiji have a reasonably well-developed service, industrial, and mining sector. Tourism is important for some Pacific countries, particularly French Polynesia, New Caledonia, Vanuatu, Fiji, and Cook Islands. Expatriate communities resident in countries such as Australia, New Zealand, and the USA send significant remittances back to their original home countries: Samoa and Tonga benefit from such remittances. A significant proportion of the population of many Pacific Island countries live traditional lives with only limited excursions into the cash economy, living instead on subsistence agricultural and fisheries livelihoods and being self-sufficient for shelter and infrastructure needs. **Figure 5** illustrates the GDP/Head in US dollars for Pacific economies. Countries which are territories (even if self-governing) or are politically part of western economies, such as French Polynesia, New Caledonia, Wallis and Futuna, Cook Islands, Guam, and Niue, have relatively high GDP/Head figures, particularly PSIDS such as French Polynesia with a GDP/Head of US$36,000 (higher than South Korea). Independent PSIDS have GDP/Heads of between US$8000 and c. US$1700. Levels of acute poverty, with people earning ≤US$2/day, are low in the Pacific Islands region. Subsistence livelihoods are difficult to account for in terms of traditional economic parameters such as GDP/Head. People have sufficient food to survive, although some diets may be limited, or lead to obesity, heart disease, and diabetes (levels of non-communicable diseases (NCDs) are high in the Pacific region). Pacific populations are small by Asian standards: The whole Pacific region has a total population of c. 11 million people. PNG is by far the most populous country with >8 million people, and other populations include Fiji (900,000), Solomon Islands (610,000), Vanuatu (264,000), Samoa (187,000), Kiribati (109,000), Tonga (103,000), and Marshall Islands (54,000).

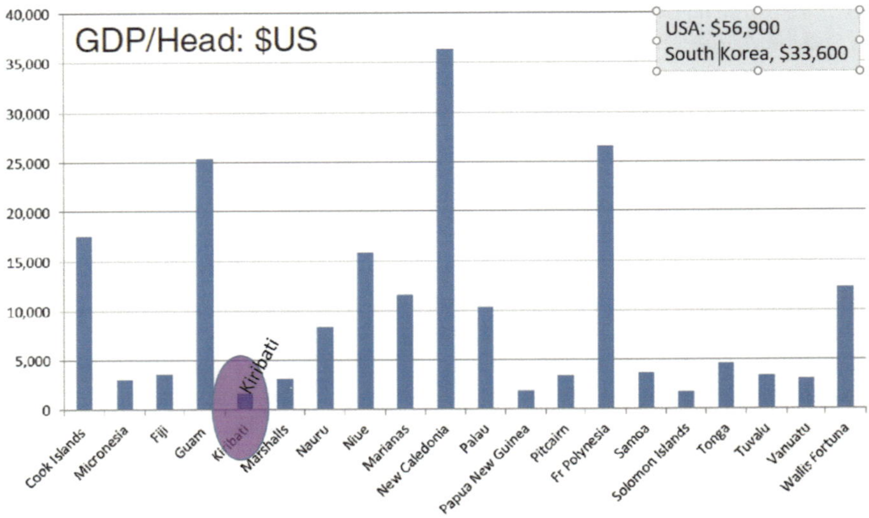

Figure 5.
Graph of GDP/head for Pacific Island countries and territories. GDP/head for Korea and the USA for reference. Note how the independent PSIDS have the lowest GDP/head values. Kiribati is highlighted (figures from SPC [15] and UN [16]).

Poverty and development challenges for the PSIDS are different to those of the poorest populations in sub-Sahara Africa, South Asia, and similar regions. PSIDS do not experience the desperate poverty of the c. 1 billion people in the world who live on <US$2 a day. The term 'persistent poverty' is sometimes applied to the lower-income populations in PSIDS which expresses a lifestyle at the limits of income/need and a situation that struggles if economic shocks impact on individuals and families (e.g. [14]). Archipelago nations can be challenging to govern and administer. The costs of providing even basic services to a widely scattered population are very high for many PSIDS. There is a trend for populations moving towards urbanised centres such as Suva, Nadi, Labasa, and Lautoka (Fiji), Honiara and Ghizo (Solomon Islands), Nuku'alofa (Tonga), and Majuro (Marshall Islands). For many PSIDS, employment opportunities are limited within the cash economy, and young people in particular, can struggle. Fertility rates are relatively high in the PSIDS region, although falling in some countries. Most PSIDS populations are youthful. Highest fertility rates (≥4 children per woman) include the Solomon Islands, Samoa, PNG, Vanuatu, and the Marshall Islands. Lowest fertility rates (2.1 children or less per woman) occur in French Polynesia and Palau, with other PSIDS somewhere between these end-member situations [18]. Life expectancies are relatively low in many PSIDS and particularly low for Kiribati, PNG, and Nauru (<59), with many other PSIDS having a life expectancy of 69 [19]. NCDs may be a contributory factor to the relatively low life expectancies within PSIDS. Many countries exhibit a classic pyramid-shaped demographic curve of age vs. percentage of population, with a high young/old population ratio (**Figure 6**).

The geographical isolation of PSIDS including distance to markets is a significant barrier to economic development. This leads to high transportation and import/export costs. Some PSIDS businesses have developed a global reach in spite of these handicaps: mineral groundwater from Fiji is a good example here. Industries that can develop independently of geography, such as the knowledge economy, and internet-based businesses may be a way forward, particularly if fast broadband develops (again geographical distances make the laying of fibre optic cables expensive). Niche agricultural and manufactured products have small market bases in

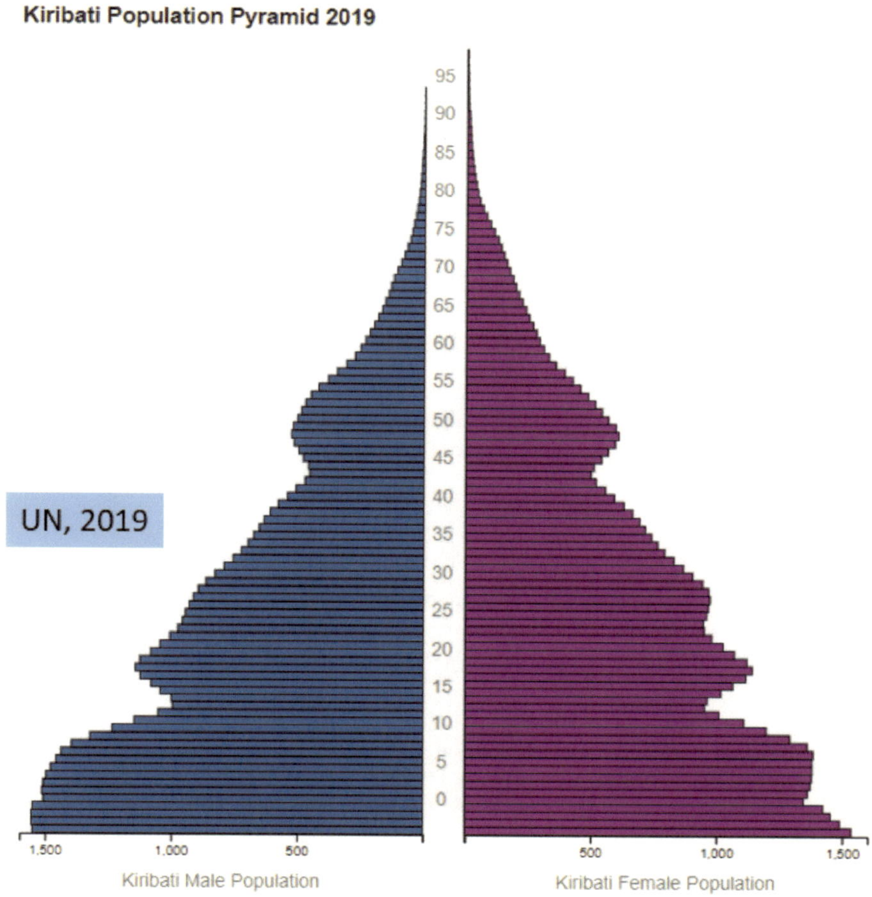

Figure 6.
Age-frequency diagram for males and females in Kiribati. Note the pyramidal form of the curve indicating a high fertility rate. 'Indentations' in the curve correspond to I-Kiribati who may study or work overseas for part of their life. The proportion of people >60 is low, and the ratio <25/>60 is very high indicating the dominance of young people within the demography. Kiribati has a fertility rate of between 2 and 4 births per woman and a life expectancy of 55–59 (acknowledgements UN [16]).

- 22 Small Island States, 16 sovereign nations
- Papua New Guinea is 'unusual' as it has a large landmass and population (460,000km2, >8 million population)
- ***Archipelago nations: many islands spread over large ocean area***
- Subsistence Economy remains strong in rural areas
- ***Limited opportunities for economic development & jobs***
- Country size varies from 28,000km2 (Solomons) to 12km2 (Tuvalu)
- Country populations vary from 900,000 (Fiji) to 1500 (Niue)
- ***Geographical isolationism***
- ***High transportation costs by sea***
- Communal culture and Indigenous peoples societal values
- ***Comprehensive, affordable, reliable, green, electricity can assist with development***

Table 1.
Summary of development challenges for the Pacific Islands region.

Australia and New Zealand. Fisheries are in high demand, and PSIDS are receiving higher returns on fisheries than in previous times. There are, however, threats to diminishing fish stocks. Small internal populations with limited spending power are an additional constraint on the development of locally based internal economies.

As a specific example, Kiribati is typical of the smaller atoll PSIDS (**Tables 1** and **2**, [17, 20]). The country is spread over three island groups (Gilbert, Line, and Phoenix) with only the Gilbert Islands being particularly populated, although Kiritimati (or Christmas Island), a second urbanised island, is situated in the northern part of the Phoenix Islands. Thirty three island groups are spread over an ocean area of 3.5 million km^2, presenting tremendous challenges for a small

- 33 island atolls, 3.5 million km2 total area, land area 810 km2
- Population 115,000 in 24 islands. >50,000 live in South Tarawa, 6,500 in Kirimati
- Primary Income: fishing, coconut products: *imports >>>exports*
- Limited availability of water supplies (ground water, rainwater harvesting)
- GDP/Capita: $1838 *(South Korea, $33,600,742, USA $56,900, Laos, $2,457, Fiji, 5,589)*
- High dependence on imported oil. 49% of imported oil for S. Tarawa power generation. This cost $6.1M US.
- 24.5GWh of electricity demand in S. Tarawa, 1.65GWh demand in Christmas Island
- Electricity demand: 41% residential, 34% government, 19% commercial
- Peak load in South Tarawa c. 3-4MW
- South Tarawa: 5.45MW of installed capacity (diesel). Solar power to add c. 1.4MW (0.5MW solar power installed in 2015, more on-stream)
- Kirimati: 1.5MW installed capacity (diesel) + solar power
- Outer Islands have smaller diesel and solar generating capacity
- Kiribati Integrated Energy Roadmap 2017-25: 45%-60% reduction in fossil fuels by 2025
- South Tarawa Renewable Energy Project to fund: 4.1MW of solar power/2.6 MW of storage

Table 2.
Summary of the development and energy context of Kiribati (UN [16], World Bank [20], NZMFAT [21], United Nations [22]).

Figure 7.
Many atoll PSIDS have developed high-density concentration urban centres which attract populations from the outer islands. These islands are characterised by high densities of housing, many of which are traditional houses and some of lower-quality informal style. Examples of urbanised centres include Funafuti (Tuvalu), South Tarawa (Kiribati), and Ebeye/Majuro (Marshall Islands).

PSIDS in terms of administration, governance, service provision, and monitoring of foreign fishing fleets. Around 24 islands are populated. The total population of Kiribati is 115,000 of which almost half live in South Tarawa and over 6000 in Kiritimati. The remaining population live on small islands within village communities and a subsistence economy. Industry and employment revolves around the government and administration, services, fishing, and coconut products, together with marine services. Many I-Kiribati males serve as mariners. Tourism is small and limited and appeals to a niche market (e.g. game fishing around Kiritimati). The total GDP/Head is c. US$ 1800 which is one of the lowest figures for PSIDS. South Tarawa is an example of an urbanised atoll which comprises high densities of population and housing: Similar centres exist in the Marshall Islands and Tuvalu (**Figures 7–9**).

Figure 8.
Typical traditional house in South Tarawa, Kiribati. Urban houses such as this comprise a thatched roof and cement lower part and floor. People may keep pigs close to the house if planning regulations permit. Note the sandy soils, tropical vegetation, and standing water (Photograph: Petterson).

Figure 9.
The enchanting attractions of atoll islands (here North Tarawa, Kiribati) include the seamless change from land to ocean. Atoll islanders are equally at home on land and in the ocean and can spend much of their day working or enjoying recreation in the shallower waters that surround their low-lying islands (Photograph: Petterson).

6. Pacific Islands region energy overview

A number of countries are considered here for context with respect to the deployment of OTEC in Kiribati. Regional data presented are taken from the United Nations Department of Economics and Social Affairs, 16 February 2016, 'Electricity Profiles' publication which is part of the 'Energy Balances and Electricity Profiles' series that documents a range of electricity statistics for all nations for 2013 and the previous 5 years and related sources [15, 19–22]. From geographical and economic considerations presented in the above sections, it is apparent that the solutions for the provision of power in the Pacific region cannot rely on gridded electricity networks alone: This is an unrealistic and costly proposition. Gridded electricity is an option for centres of higher population such as Apia, Honiara, South Tarawa, and Funafuti. It is also a good solution for islands with numerous population centres within close geographic proximity on one island such as Viti Levu and Vanua Levu in Fiji. More distributed, scattered, and remote populations require a range of electricity solutions, including local village grids or small stand-alone grids for a few neighbouring houses. At present many villagers rely on discrete diesel generators, with solar energy providing alternative solutions for remote rural communities.

Electricity generation will rely far more on renewable energy technologies, rather than traditional fossil fuel-powered and centralised electricity systems, as we look to the future. Solar energy technologies have dropped exponentially in price over the past 10–20 years, and battery life now extends to over 8 h or so (e.g. [23]). This has the consequence that solar energy is now becoming an increasingly attractive option for low-power-intensity-consumption solutions in remote Pacific Island locations, as well as supplementing power supply in urban regions. Hydropower will become increasingly important. Pacific countries such as Fiji, PNG, and Samoa all utilise hydropower to a significant extent already as part of their overall energy mix (in Fiji hydropower already contributes 60% of the installed energy capacity). In the future a range of scales of hydropower from mega to micro will be employed for solutions in different geographical and social settings. Wind energy has hardly been realised in the Pacific, although a few countries have invested in small wind farms (Fiji and Vanuatu as examples). The usage of wind energy will undoubtedly grow with time, as it has in Europe over the past decade, for example. Other options include biomass-generated power plants (as tropical countries have rapid biomass growth rates: One biomass plant will shortly come on line in Fiji, developed by South Korea) and wave and tide energy (SPC has been involved in the waves and coasts in the Pacific (WACOP) project funded by the European Union, which has recommended that countries with a mean wave energy flux in excess of 7kw/m of wavelength have a particularly high potential, e.g. New Caledonia, Tonga, Cook Islands, and countries south of latitude 20° south [24]). Geothermal energy sits alongside possible options for electricity generation in Pacific countries.

Access to electricity is highly variable, and in many parts of Melanesia (e.g., PNG, Vanuatu, Solomon Islands), 60% to c. 90% of the population does not have access to electricity, with the exception of local diesel generators/solar energy. Specific national figures for percentage of population without access to electricity are Tonga (5%), Samoa (5%), Fiji (25%), Kiribati (55%), Solomon Islands (85%), and PNG (87%) [23].

Figure 10 presents a global view of GDP/capita vs. electricity usage. Once GDP/capita rises above around US$15–18,000 per capita, utilisation of energy is consistently high (c. 8–10,000 kilowatt hours per person). At lower levels of GDP/capita ($10,000 US or less), electricity utilisation is at much lower levels and is highly variable (e.g. compare a GDP/capita of c. $3000 US and electricity/capita utilisation

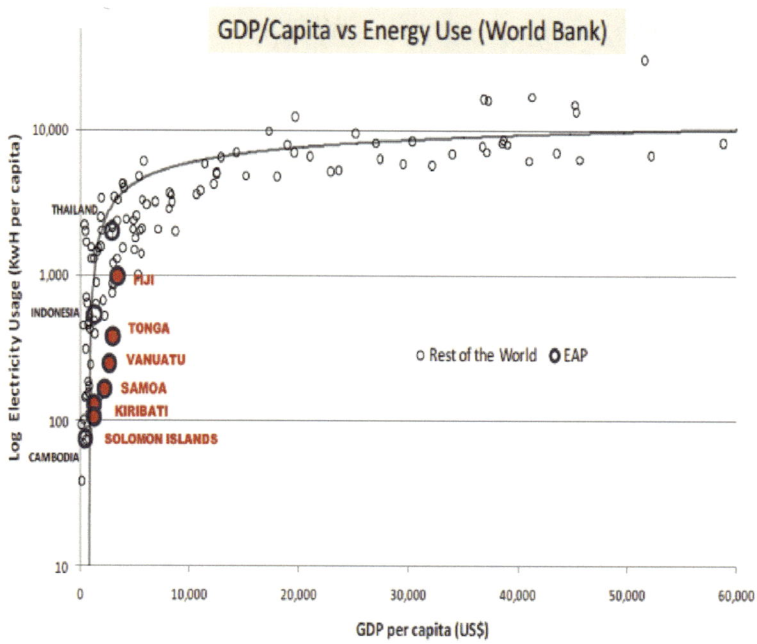

Figure 10.
Graph of GDP/capita vs. electricity usage per capita. See text for details [23].

of c. 4000 kilowatt hours/capita for Thailand to a GDP/Head of <$1000 US and electricity utilisation/capita of c. 80 kilowatt hours/capita for Cambodia). Many Pacific Island countries plot within the lower part of the global curve with a GDP/capita of between c. US$1500 and $5000 and electricity utilisation between 70 and 1000 kilowatt hours/capita. The evidence suggests that Pacific Island countries are, in the main, at the lower to lowest end of global development, when it comes to electricity utilisation per capita.

Figure 11 presents installed electricity generation capacity of selected Pacific Island countries. Papua New Guinea, with its larger population and global mining industry, generates the highest amount of electricity, with an installed electricity capacity of 827 MW, with Fiji at 255 MW, Samoa at 42 MW, and down to Kiribati, with only c. 6 MW of installed generation capacity. These are extremely low levels of installed electricity capacity by world standards. By comparison, note the installed generation capacities of China, 1.3 million MW; the USA, c. 1 M MW; India, c. 300,000 MW; the UK, 92,000 MW; Thailand, 48,000 MW; and Iceland (with a Pacific-like island population of c. 332,000) 3000 MW [16].

Figure 12 presents a more detailed analysis of installed generating capacity/Head vs. GDP/Head. What is apparent from this graph is that countries such as Fiji, Samoa, and the Marshall Islands have significantly higher installed generation capacity with respect to the strength of their economy, than countries such as Kiribati and the Solomon Islands in particular. The evidence suggests that most Pacific Islands are 'under-energised' at the present time. Some PSIDS are the most under-energised countries in the world. This lack of access to electricity is a serious inhibitor of economic social development.

Table 2 summarises a number of characteristics of the capital of Kiribati (South Tarawa) present-day energy situation [20]. South Tarawa has 5.45 MW of installed capacity provided by an ageing diesel generator which experiences regular periods of non-transmission. The diesel capacity is supplemented by solar energy: Up to an

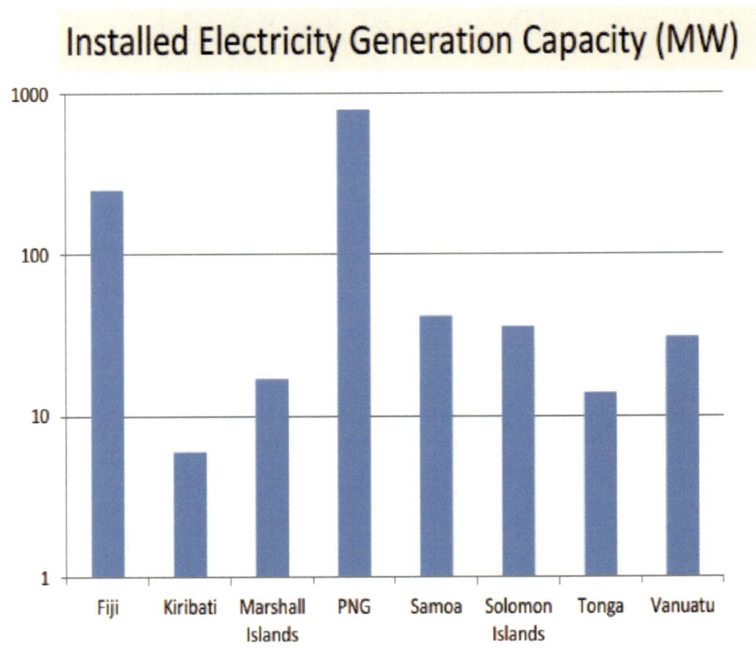

Figure 11.
Installed electricity generation for selected Pacific Island countries (data, United Nations [22]). Note the logarithmic scale on the Y-axis.

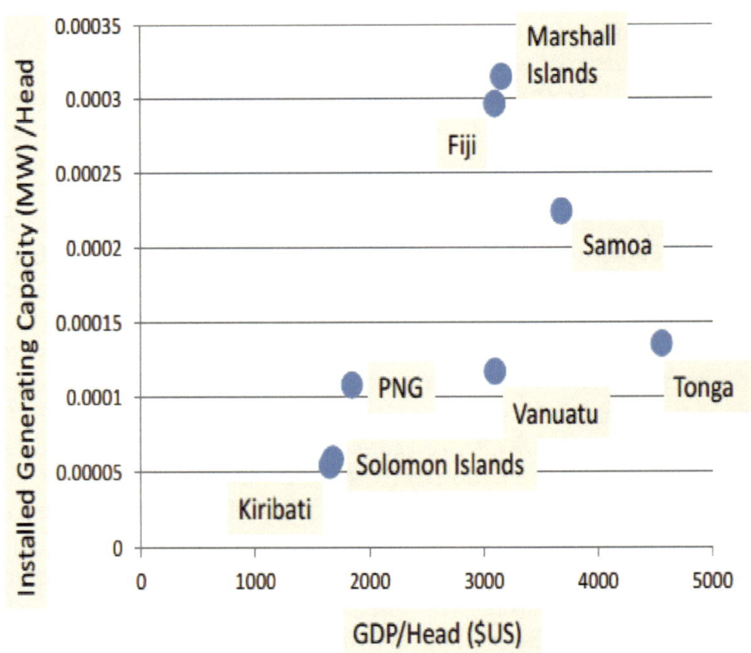

Figure 12.
Graph of installed electricity capacity per head versus GDP/head for selected Pacific Island countries. Note how Kiribati and Solomon Islands are the least energized countries and Fiji/Marshall Islands the most energized from this analysis.

additional 1.5 MW is on line or planned for the future. Peak demand in South Tarawa is around 3–4 MW. Kiritimati has 1.5 MW of installed capacity with smaller stand-alone diesel providers in the smaller islands. There is a Kiribati roadmap for energy which aims to move towards a less fossil fuel- and imported oil-reliant future, with solar power (and OTEC) planned to be the main renewable energy providers for the future. The 2017–2025 Kiribati energy roadmap plans for a c. 45–60% cut in fossil fuel generated energy by 2025.

7. OTEC and global OTEC resources

Figure 13 shows the world map of realisable ocean thermal energy resources, with the greatest temperature gradients (from surface to water depths of 1 km) shown in red and the minimal possible realisable OTEC temperature gradient shown in blue. Ongoing and planned OTEC projects are named. A minimum temperature differential of 17–20°C between surface waters and those at 1 km depth is required for OTEC operations. This condition is met, year-round, for tropical and subtropical waters in all oceans. Some areas at the fringes of the ocean thermal resource, such as Japan, South Korea, and the Arabian Peninsula, have the capacity to generate power through OTEC for part of the year. Kiribati and its capital township, South Tarawa, are situated within an ocean area with the highest thermal difference between surface waters and waters at 1 km depth (c. 24–28°C).

Figure 14 shows the principles of OTEC technology. In a closed cycle OTEC system, such as will be deployed on Kiribati, a working fluid of R32 (seawater can be used as working fluid in open cycle OTEC systems, e.g. Hawaii) is vaporised, with the vapour driving a turbine to create electricity. The vapour is condensed by heat exchange with colder ocean water and then heated/vaporised via heat exchange with warmer waters, and the cycle continues. OTEC boasts little to no seasonal variation throughout the day and seasons. For remote islands and coastal villages that have no power grids, OTEC can provide clean, self-reliant, sustainable energy.

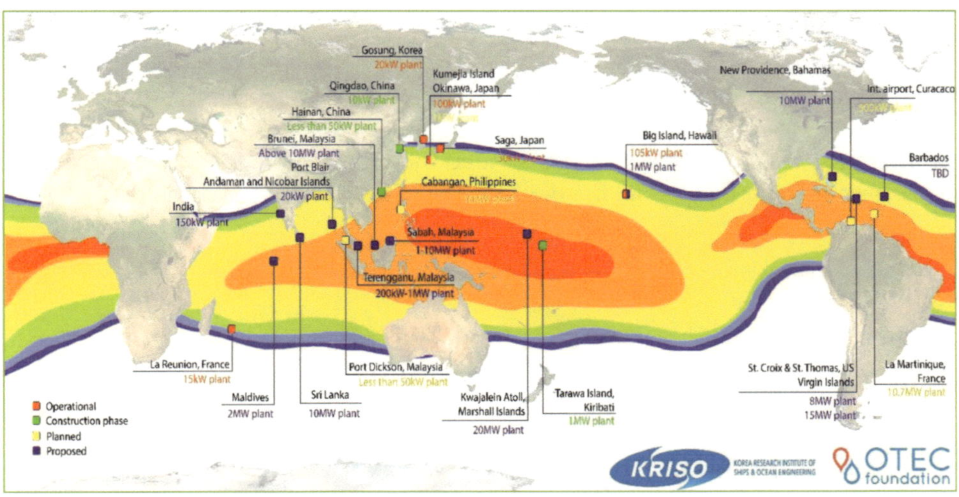

Figure 13.
Global map of OTEC activities and resource in terms of the temperature difference between surface seawater and seawater at a depth of 1 km. The highest temperatures (and highest potential OTEC energy resources) are situated NE and E of Papua New Guinea, Indonesia, and the Philippines. Significant thermal resources are present within tropical and subtropical waters in all oceans and can benefit SIDS and continental countries within this area. Kiribati and its capital township of South Tarawa lie within the 'bulls eye' of thermal energy resources. A minimum temperature difference of 17°C between surface waters and waters at 1 km depth are required for OTEC at the present time (acknowledgements KRISO).

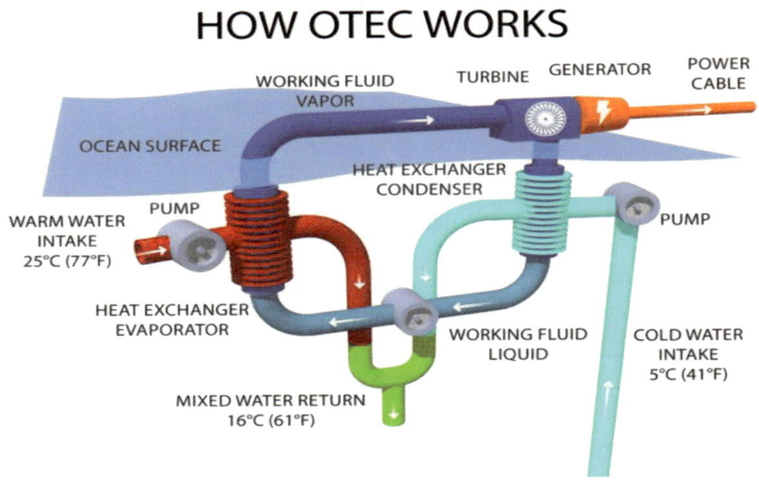

Figure 14.
Principles of OTEC. A working fluid (R32 within closed cycle OTEC plant such as on Kiribati) is vaporised, with the vapour turning a turbine to create electricity. The vapour is cooled from deeper seawater and then heated via heat exchanges to be vaporised once more. OTEC plants can also provide desalinated drinking water and waters for agriculture/aquaculture at downstream (acknowledgements Scientific American [25]).

8. Design and fabrication of the KRISO 1 MW OTEC demonstration plant

As part of the development of the KRISO 1 MW plant, a closed cycle OTEC system was initially designed and implemented. Closed cycles require a working fluid to transport and exchange heat within the system. The selected working fluid for the KRISO example was R32 (difluoromethane) which has a relatively high heat transfer coefficient and low environmental impact. Typical environmental characteristics of R32 include a relatively low ozone depletion index (ODP) and global warming potential (GWP) of 0 and 675, respectively. Various studies have been conducted to improve the performance of OTEC cycles, utilising multistage cycles, Kalina cycles, Uehara cycles, and so on [5, 6]. For the KRISO case, a simple Rankine cycle was applied to demonstrate the long-term operational potential of the OTEC plant. **Figure 15** shows the experimental temperature-entropy (heat transfer divided by the temperature) performance behaviour of the OTEC cycle, for the KRISO-OTEC plant, with field conditions of

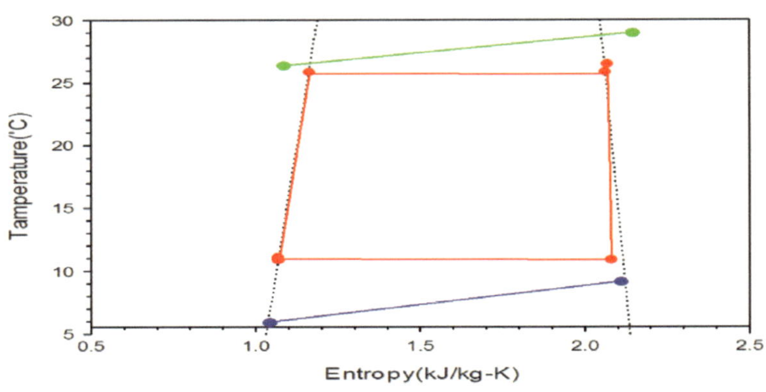

Figure 15.
Temperature-entropy (heat transfer divided by the temperature). Diagram of an OTEC cycle (after [5, 6]).

29°C surface seawater temperatures and 5°C deeper seawater temperatures. Closed cycles are circulated by the working fluid pump, so an endless cycle of power is formed when a heat source is supplied. The turbine output (Wt) of a closed cycle OTEC plant is given by Eq. (1). The net power generation (Wnet) is given by Eq. (2). The cycle was designed using Aspen HYSYS ver8.0, a process design program [5, 6].

$$W_t = m_r \bullet (h_{ti} - h_{to}) = m_r \bullet \eta_t \bullet (h_{ti} - h_{tos}) \tag{1}$$

$$W_{net} = W_t - w_{wwp} - w_{cwp} - w_{rp} \tag{2}$$

where mr is the mass flow rate of the refrigerant, ht_i and ht_o are the inlet and outlet enthalpy of the turbine, ηt is the efficiency of the turbine, and h_{tos} is the turbine outlet enthalpy of the isentropic process. $\omega_{\omega\omega p}$, $\omega_{c\omega p}$, and ω_{rp} represent the pump output of surface water, deep water, and working fluid, respectively.

The heat of evaporation of temperature difference generation is shown in Eq. (3), and the net power efficiency (η_{net}) is calculated by the ratio of the amount of net power generation (Wnet) and evaporation heat (Qω) as in Eq. (4) [5, 6].

$$Q_w = G_h \bullet C_h \bullet \Delta T_{in-out} \tag{3}$$

$$\eta_{net} = W_{net} / Q_w \tag{4}$$

where G_h is the surface water flow rate, C_h is the specific heat of seawater, and ΔTin-out is the temperature difference between the evaporator inlet and outlet (**Figure 14**).

Parameter	Value	Unit
Hot water inlet temperature	29	°C
Hot water mass flow rate	1948.5	kg/s
Cold water inlet temperature	5	°C
Cold water mass flow rate	1805	kg/s
Sea water pump efficiency	80	%
Hot water pump power	130.5	kW
Cold water pump power	188.3	kW
Heat source capacity	32,364	kW
Heat sink capacity	31,148	kW
Refrigerant pump power	82.63	kW
Turbine inlet pressure	1729	kPa
Turbine inlet temperature	26.5	°C
Turbine efficiency	85	%
Gross power	1278	kW
System efficiency	3.95	%
Net system power	876.57	kW
Net system efficiency	2.71	%

Table 3.
Analysis result of Rankine cycle OTEC demonstration plant.

Although machine efficiency conditions are dependent upon on the temperature and flow rate of surface and deep seawaters, it is assumed that the OTEC cycle operation satisfies the heat exchanger pinch temperature of 1.5°C. Results from the earlier KRISO-OTEC experiments indicated that at an 85% turbine efficiency, the power generation was 1278 kW and the new power efficiency was 2.71% (calculated using principles and equations in [5, 6]) (**Table 3**).

Based on earlier experimental results and experience, a new 1 MW OTEC plant, named *K-OTEC1000*, was designed, and core devices such as a turbine generator, condenser, and evaporator were manufactured from 2016 to the first half of 2019. The *K-OTEC1000* plant was installed on a barge ship and became a de facto offshore power plant for short-term experiments in the seas offshore from Busan, South Korea. When *K-OTEC1000* is installed on South Tarawa, Kiribati, it will be a land-based *onshore* power plant (see **Figure 16** for an artist impression of the plant).

A field experiment was conducted in the southern sea of South Korea's East Sea, offshore of Busan, to verify the OTEC plant *K-OTEC1000* performance

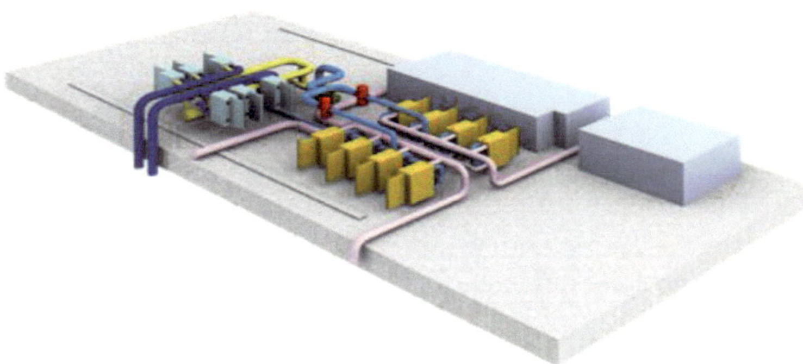

Figure 16.
3D model/artist impression of the Kiribati-based KRISO 1 MW OTEC plant.

Deployment Stage	Variable	Wave		Tidal		OTEC	
		Min	Max[1]	Min	Max	Min	Max
First array / First Project[2]	Project Capacity (MW)	1	3[3]	0.3	10	0.1	5
	CAPEX ($/kW)	4000	18100	5100	14600	25000	45000
	OPEX ($/kW per year)	140	1500	160	1160	800	1440
Second array/ Second Project	Project Capacity (MW)	1	10	0.5	28	10	20
	CAPEX ($/kW)	3600	15300	4300	8700	15000	30000
	OPEX ($/kW per year)	100	500	150	530	480	950
	Availability (%)	85%	98%	85%	98%	95%	95%
	Capacity Factor (%)	30%	35%	35%	42%	97%	97%
	LCOE ($/MWh)	210	670	210	470	350	650
First Commercial-scale Project	Project Capacity (MW)	2	75	3	90	100	100
	CAPEX ($/kW)	2700	9100	3300	5600	7000	13000
	OPEX ($/kW per year)	70	380	90	400	340	620
	Availability (%)	95%	98%	92%	98%	95%	95%
	Capacity Factor (%)	35%	40%	35%	40%	97%	97%
	LCOE ($/MWh)	120	470	130	280	150	280

Table 4.
Cost analysis in terms of capital expenditure (CAPEX) operational expenditure (OPEX), year-round availability/capacity, and the levelised cost equivalent (LCOE) (taking into account plant lifecycle production and operational costs) of OTEC power plant with other ocean energy systems [26].

characteristics, in field conditions, prior to transport and installation on Kiribati. The experiment was conducted at the end of September, 2019, and proved to be successful, yielding a significant amount of operational data and proving the design concept [4]. During the experiment, the surface water temperature decreased, and the temperature difference between the surface and the deep water decreased to 18.7°C, with the output power level at 338 kW. This was a lower output than hoped for, but field conditions were suboptimal as the Korean winter was rapidly approaching. The data suggested however that the *K-OTEC1000* plant could likely produce 1 MW at a temperature differential of 24°C and 500 kW at a temperature differential of 20°C. The experiment produced the highest ever energy output from a seaborne OTEC plant, in the world, to date, setting a new world record [4].

The usefulness of OTEC-generated electricity to Kiribati and other islands will depend upon social acceptability and economic feasibility. As can be seen in **Table 4** [26], the unit production cost of electricity from the OTEC plant is estimated to be US$ 0.15 ~ 0.65/kWh. Currently, electricity costs in Kiribati are between US$ 0.57 and 1.59/kWh. These data suggest that OTEC could be commercially competitive and viable for a Pacific Island situation. Because of Kiribati's favourable position in equatorial waters, it will be possible for OTEC plants to operate year-round on a 24/7 basis. If the OTEC plant proves successful, it could be scaled up to a 10 MW plant which would provide all of South Tarawa's current electricity requirements with no concomitant production of Greenhouse Gases.

9. Kiribati and OTEC deployment

KRISO plans to deploy a 1 MW land-based OTEC (K-OTEC1000) plant for 1 year in South Tarawa, Kiribati, in 2020 to 2021. If tests are successful, this may lead to longer-term projects and perhaps fully ocean-deployed OTEC systems. The OTEC plant will be located in the Eastern part of South Tarawa (**Figure 21**). This part of South Tarawa exhibits a rapid bathymetric gradient, reaching ocean depths of c. 3.5–4 km ocean depth within a 5-km horizontal distance (SOPAC [27]). This location allows land-based OTEC plants to access c. 1 km deep waters via pipes which hug the offshore underwater slopes and can be cemented and secured onto a solid rock foundation. Plans are now being made to transport the OTEC plant (**Figures 16–20**), by ship, from Busan to South Tarawa, in 2020/21. All environmental and planning permissions were granted by the Kiribati government in 2018. Ocean physical and chemical parameters have been measured in South Tarawa for a number of years, and the KRISO team have a good working knowledge of variations in temperature, pH, salinity, and redox conditions in the lagoon, and ocean, close to the proposed OTEC plant site.

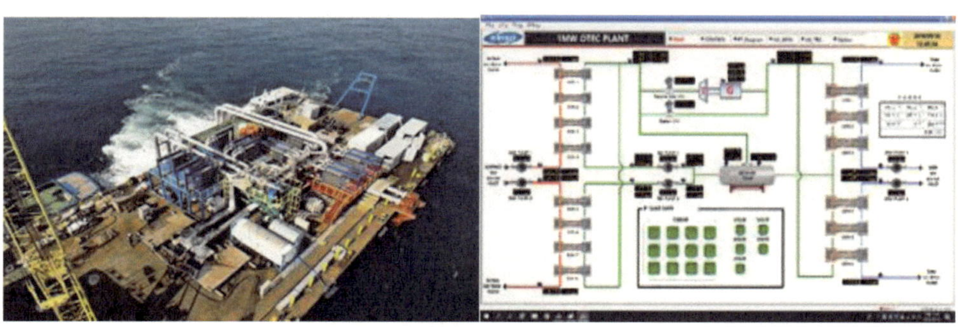

Figure 17.
Operation scene of barge-mounted 1 MW OTEC plant (L) and monitoring system (R).

Once the OTEC plant is deployed, it has funding and permissions to test the equipment in 2020/2021. During this time the whole programme will be assessing applications for the South Tarawa community for electricity and downstream utilisation for drinking water and agriculture/aquaculture. I-Kiribati people will be trained in OTEC-related engineering and science and the application of products for agriculture and aquaculture. If the project proves successful, the KRISO team will be making applications for funding for longer-term OTEC deployment on Kiribati, and for scaling up the 1 MW plant, with ambitions for a fully ocean-deployed OTEC plant generating 5–50 MW of electricity in the future (**Figures 21** and **22**).

Figure 18.
Key components of 1 MW OTEC plant of K-OTEC1000, which was loaded onto a barge ship for experimental tests, offshore from Busan, South Korea.

Figure 19.
Perspective view of floating OTEC plant depicted by KRISO.

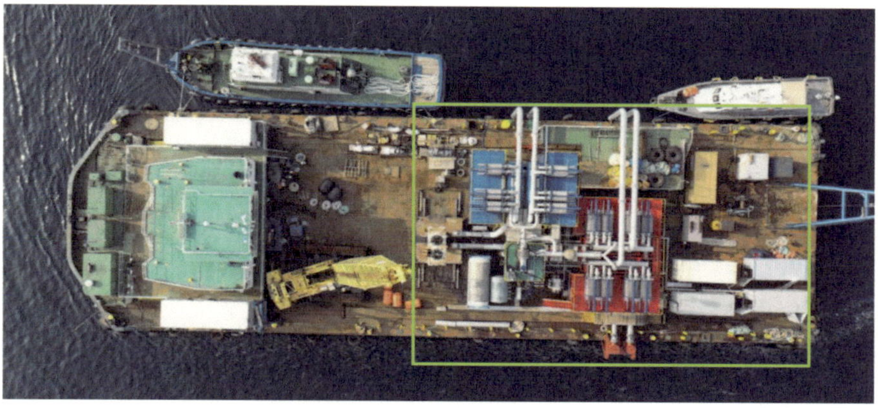

Figure 20.
K-OTEC1000 plant onboard a barge ship in the eastern seas, near Busan, South Korea. The full extent of the OTEC plant is shown within the box in the figure. The plant was successfully tested in September, 2019, and it will be transported to South Tarawa in 2020 (Photograph: Kim).

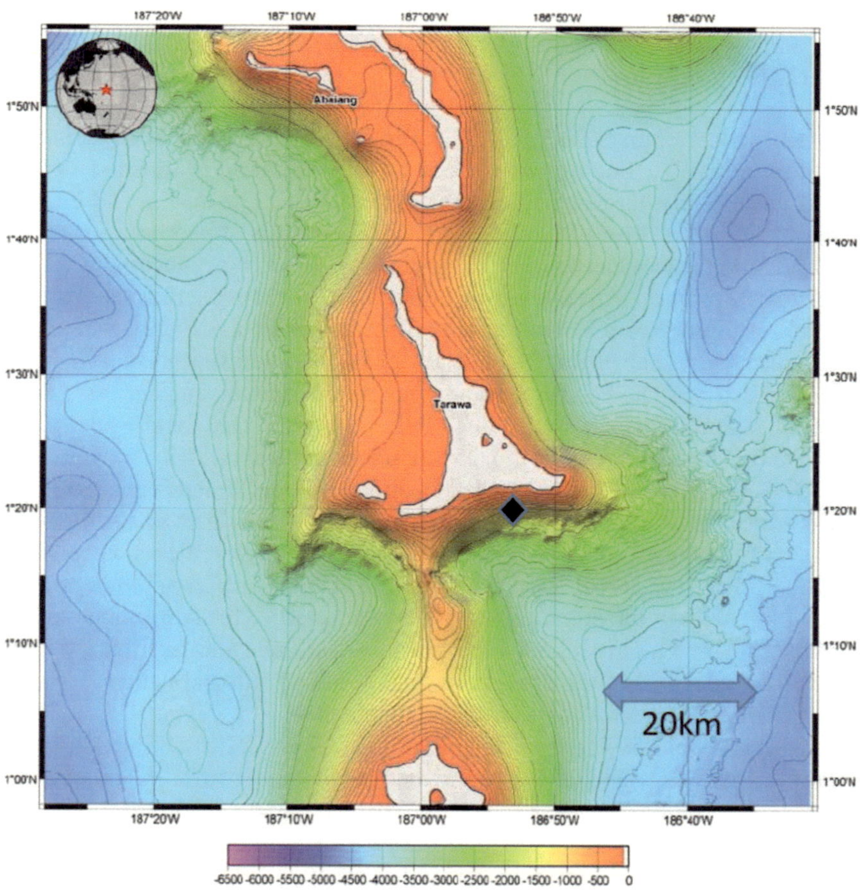

Figure 21.
Bathymetric map of South Tarawa showing the probable location of the 1 MW OTEC plant (black diamond). Note the rapid drop-off in depth away from the atoll allowing an OTEC plant ready-access to deep water and the market of South Tarawa. South Tarawa is a seismically quiet area with extreme storm events occurring relatively infrequently and quiet seawater conditions (acknowledgements, SOPAC [27]).

Figure 22.
Bird's eye view/artists impression of the building for sustainable seawater utilisation Center (SSUC). Downstream utilisation of discharged seawater for district air conditioning, desalination, aquaculture, and agriculture applications will be delivered for capacity building and SDGs achievement in Kiribati and coastal communities along the tropical belt (acknowledgements, KRISO).

10. Concluding statements

The deployment of an OTEC plant in South Tarawa is ambitious in technical and development terms. Technologically, the South Tarawa plant is destined, at least for a while, to become the largest OTEC plant in the world. The design and building of the plant represents a US$20 M investment and thousands of people-hours of highly expert time. The project has been ongoing since 2013, with predecessor OTEC laboratory-scale experiments extending back to 2010. Transporting the main plant and c. 3 km of piping from Korea to South Tarawa will be a difficult challenge.

The final deployment of the OTEC plant in Kiribati will represent a high water mark for Kiribati and the Pacific region in general. Infamous nuclear tests aside (in the Marshall Islands, Kiribati, and French Polynesia), it is rare for the Pacific Islands region to be the centre of major global-scale scientific-technological advances.

The KRISO-OTEC-Pacific project has been carefully planned from 2013. KRISO and the Korean government are to be commended for the effort and time invested in discussing the programme, awareness raising, and genuine inclusive reciprocal Pacific Islander activities that have predated the OTEC Kiribati deployment. It is challenging for widely different cultures to work together for common goals with development goals. The model here that combines science and technology with a developmental goal within the context of Pacific Island cultural values is commendable from an *interconnected geoscience* perspective.

The project, should, in theory bring benefits all-round. KRISO, the South Korean Government, and the world OTEC community will benefit from the on-site, 1-year technical testing and operation of a land-based 1 MW plant. If this is successful, it may lead to more ambitious, larger-scale OTEC developments, including a full-ocean OTEC plant. The project may encourage other OTEC workers around the world, in the Caribbean, Indian Ocean islands, and beyond, to accelerate their OTEC developments.

If, however, all that results from the OTEC Kiribati experience are technological benefits, this will be a disappointment.

As this paper has analysed, small atoll PSIDS and SIDS in general are in acute need of many of the benefits OTEC technologies can bring. A lack of affordable, reliable, and accessible electricity is a challenging constraint on development, anywhere in the world. It points to a chicken and egg situation: Minimal power results in minimal development, and limited development results in limited power infrastructure being created. Industry that is attracted to Kiribati, such as tuna canning factories, have to develop much of the infrastructure they require, such as power and water, themselves. This adds significantly to already-high start-up costs and is a major disincentive to inward investment into Kiribati and other SIDS. If the state can provide fundamental infrastructure, including reliable, affordable electricity and water supplies, conditions for inward investment are improved.

OTEC not only brings electricity but the promise of associated seawater technologies and industry. Deep ocean water has proven chemical and biological qualities that can be applied to products and services for human health, cosmetics, agriculture, and aquaculture. In Goseong, South Korea, next to the OTEC and seawater utilisation plant, a number of high-technology industries have developed that produce mineral salts, mineral waters, and cosmetics. These could develop, even in an unlikely setting such as South Tarawa, given training, investment, and the establishment of a Sustainable Seawater Utilisation Center (SSUC). Waters linked to OTEC operation can be used to develop hydroponic agricultural plants, refrigeration plants, and aquaculture plants for shellfish and fish farms. These developments can all contribute to many sustainable development goals.

Alongside science and technology, and the development of agriculture and industry, with related high-quality employment opportunities, is the potential for new education and training. KRISO is working with the Kiribati government in helping develop a science and technology of sustainable seawater utilisation at the centre. The presence of a fully operational and long-term OTEC plant would be a catalyst and encourager for the further development of these much-needed educational initiatives for Kiribati, which can then become a developmental model for other PSIDS and SIDS in general.

The 1-year testing of a land-based 1 MW OTEC plant in Kiribati in 2020/2021 and the 7-year lead up to this deployment is highly innovative and a good model for international development. Time will tell if the 1-year duration test develops into something far more significant from a Kiribati development perspective.

Acknowledgements

This research was supported by a grant from National R&D Project of 'Development of 1MW Ocean Thermal Energy Conversion Plant for Demonstration' (PMS4080, PMS4320) funded by the Ministry of Oceans and Fisheries, Republic of Korea. Auckland University of Technology (AUT) are acknowledged for their support to the lead author.

Author details

Michael G. Petterson[1*] and Hyeon Ju Kim[2]

1 School of Science, Auckland University of Technology, Auckland, New Zealand

2 Seawater Energy Plant Research Center, Korea Research Institute of Ships and Ocean Engineering, Goseong-Gun, Gangwon-Do, South Korea

*Address all correspondence to: michael.petterson@aut.ac.nz

References

[1] United Nations Paris Agreement. 2015. Available from: https://unfccc.int/sites/default/files/english_paris_agreement.pdf

[2] Petterson MG. Interconnected Geoscience for International Development. Episodes Journal of International Geoscience, International Union of Geological Sciences. 2019; **42**(3):225-233. DOI: 10.18814/epiiugs/2019/019018

[3] United Nations. Resolution A/RES/70/1. Transforming our world: The 2030 agenda for sustainable development. Records of the Seventieth Session of the General Assembly of the United Nations, Agenda Items 15 and 116; 21 October, 2015. p. 35

[4] Lee HS, Yoon JI, Son CH, Ha SJ, Seol SH, Ye BH, et al. Efficiency enhancement of the ocean thermal energy conversion system with a water-vapor ejector. Advances in Mechanical Engineering. 2015;**7**(3). DOI: 10.1177/1687814015571036

[5] Kim H-J et al. Demonstration of 1MW OTEC plant in Korean waters in 2019. In: Proceedings of the 7th International OTEC Symposium; 2019. p. 5

[6] Lim S-T, Kim H-J, Lee H-S. Dynamic simulation of performance change of MW-classOTEC according to seawater flow rate. Journal of the Korean Society for Power System Engineering. 2019; **23**(1):48-56

[7] Lim S-T, Lee H-S, Kim H-J. Simulation of power generation performance of MW-class open- and closed-cycle OTEC systems based on seawater temperature change. Journal of the Korean Society of Marine Engineering. 2019;**43**(6):420-426

[8] Lee DY, Nam BW, Hong SY, Kim H-J. An experimental study on the motion response of OTEC platforms. Korea Research Institute of Ships and Ocean Engineerng (KRISO). In: Proceedings of the 25th International Ocean and Polar Engineering Conference. Kona, Big Island, Hawaii, USA. 21–26 June, 2015

[9] Brundtland G. Report of the World Commission on Environment and Development: Our common future. United Nations General Assembly Document No. A/42/427; 1987

[10] Gill JC, Bullough F. Geoscience engagement in global development frameworks. Annals of Geophysics. 2017;**60**:1-10. DOI: 10.4401/ag-7460

[11] Stewart IS, Gill JC. Social geology—Integrating sustainability concepts into Earth sciences. Proceedings of the Geologists Association. 2017. DOI: 10.1016/j.pgeola.2017.01.002

[12] Stewart IS. Sustainable geoscience. Nature Geoscience. 2016;**9**(4):262

[13] CEBR. World Economic League Table, 2019. A World Economic League Table with Forecasts for 193 Countries to 2033. London, UK: Centre for Economics Business Research Ltd.; 2019. p. 214

[14] Petterson MG, Tawake AK. Toward inclusive development of the Pacific region using geoscience. In: Wessel GR, Greenberg JK, editors. Geoscience for the Public Good and Global Development: Toward a Sustainable Future. Geological Society of America Special Paper. Vol. 520. 2016. pp. 459-478

[15] SPC. Pocket Statistical Summary. Noumea, New Caledonia. 2013. p. 4. Available from: www.spc.int/prism

[16] UN. Gross National Product per Head Statistical Data. 2019. Available from: http://data.un.org/Search.aspx?q=GDP+per+capita

[17] UNDP. Human Development Indices and Indicators: 2018 Statistical Update. Kiribati. 2019. p. 9. Available from: http://hdr.undp.org/en/data

[18] SPC. Fertility Rates in Pacific Island Countries and Territories. New Caledonia: Noumea; 2019. p. 99. ISBN: 978–982–00-1203-5

[19] SPC. Mortality Trends in Pacific Island States. New Caledonia: Noumea; 2014. p. 81

[20] World Bank. Implementation Completion and Results Report TFOA5646 on a Small Grant in the Amount of US$ 0.3 Million to the Republic of Kiribati for the Scaling-up of Renewable Energy Program Investment Plan. Report No 138090-KI. Washington DC, USA; 2019. p. 112

[21] NZMFAT. Pacific Energy Country Profiles. Wellington, New Zealand: New Zealand Ministry of Foreign Affairs and Trade; 2016. p. 21. ISBN: 978–0–477-10250-6

[22] United Nations. Electricity Profiles: Energy Balances and Electricity Profile. Department of Economics and Social Affairs; 2016. ISBN: 9789211616088. https://unstats.un.org/unsd/energystats/pubs/balance/

[23] Asian Development Bank. Vanuatu National Energy Road Map. Port Vila, Vanuatu: Government of Vanuatu; 2013. p. 99

[24] Bosserelle C, Sandeep R, Kruger J. Waves and Coasts in the Pacific: Cost Analysis of Wave Energy in the Pacific. Publication of the Pacific Community. Suva, Fiji: SPC Publications; 2016. p. 45. ISBN: 978-982-00-094-8. Available from: www.spc.int/gsd/wacop

[25] Scientific American. Hawaii First to Harness Deep-Ocean Temperatures for Power. 2015. Available from: https://www.scientificamerican.com/article/hawaii-first-to-harness-deep-ocean-temperatures-for-power/

[26] IEA-OES. International Levelised Cost of Energy for Ocean Energy Technologies. 2015. p. 48

[27] SOPAC, Sharma A, Kruger J. Kiribati Technical Report. High Resolution Bathymetric Survey in Kiribati. SOPAC Project Report No 114. Suva, Fiji; 2008. p. 44

Chapter 2

The Social Energy: Contexts for Its Assessment

Alonso Pérez Pérez, Mauricio Latapí Agudelo and Graciela Rivera Camacho

Abstract

The recognition that interdisciplinary approaches must be integrated in any planning that concerns science and technology has shifted from theoretical perspectives to practical implementation. The once distant worlds of Engineering and Social Sciences must cross their boundaries if they are to contribute to the transition toward a carbon-neutral energy supply. The objective of this chapter is to explicitly state the connection of the social dimension with engineering applications (such as the OTEC technology) by developing basic elements of social analysis and explaining the conceptual and practical framework of instruments and practices like the social impact assessment. The chapter will mention the holistic perspective of the Sustainable Development Goals (SDGs), and the general debate of integrating the social sciences and humanities in environmental change research.

Keywords: social dimension, interdiscipline, social impact assessment, Sustainable Development Goals

1. Introduction

Imagine, as an engineer, a day in the university. Once you have found your classroom and taken a seat, the professor starts to write in the chalkboard: Joseph Fourier. So far, nothing strange. You recall your thermal engineering lessons and therefore the transport equations where the surname Fourier became familiar for the first time. You grab your pen as the professor starts to talk.

-Lets recapitulate our last theme. We saw the important role Fourier played in...

You write in your notebook "Fourier's important role in..."

-...in the public commissions on insurance and social statistics in France, as the director of Recherches statistiques sur la ville de Paris et le départment de la Seine between 1826 and 1829.

What a confusion! For sure you have mistakenly entered a Social Science class. As you prepare to leave the room, the professor goes on.

-Remember how Laplace established calculations of the probability that a quantity lay within certain limits? Well, Fourier took that principle—first used in the analysis

of the voting procedures of some juries in Europe—and extended its application to mass social phenomena. Just as Laplace found profoundly terrifying that the judicial system of his country executed people with a 30% chance of error, Fourier was committed to a sociopolitical project (the Science Morale of the XIX century), where a comprehensive theory about the way humans relate to each other was the goal.

Mathematics, physics, law, statistics, society, politics... All in one. Why? How come? Returning your notebook to the bench where you sat initially, you decide to stay.

By no means this chapter will talk about Fourier's biography nor Laplace's. An excellent book about the history of probability is available [1]. Instead, some of the main ideas regarding what is expected to be a social analysis (in the context of renewable energy transition) will be exposed, using an imaginary conversation between an Engineering student (ES) and a professor (P). The title refers to "social energy" as a guiding analogy to the physical one. The metaphor (in the sense of an equivalence resulting in work or heat) helps to suggest an intrinsic force that operates within societies, fundamentally, in a moral realm: not what is the case (as in Natural Sciences) but what should or should not be the case. This social guidance, translated in the scenario formulated by Sustainable Development Goals, points to a radical proposal: "While science and technology will undoubtedly play a key role in sustainability transitions, how societies choose to construct and pursue visions of sustainability will be an intensely social, political and cultural process" [2].

As the text develops, it will be mentioned how the notion of interdisciplinarity made its way through the actual perspectives of fully understanding society and nature, followed by some considerations about the stages and characteristics of project planning. At this point, a brief consideration about the interaction between theory and practice will be given in order to make sense about the role that social analysis is supposed to play in engineering applications such as OTEC technology. The Social Impact Assessment tool will be presented in an historical and actual context. Finally, all these themes will find a closure within the Sustainable Development Goals—based in scientific diagnosis and coordinated social management—which blueprint solutions to challenges facing poverty, inequality, environmental degradation and prosperity.

2. The social energy

-[ES] Professor, why does the majority of the scientists of the past knew about everything? I did not have a clue that Fourier, for example, contributed to social statistics.

-[P] That's an interesting question. Imagine that in Ancient Greece one of the most perdurable traditions was that of verbal arguing. It was a form of combat between two people who dispute the strongest knowledge about something. The rules were simple: the questioning challenged his opponent by asking a question that presented two options to solve a contradiction. As the interrogate chose one (the thesis of the argument), he bet for the truth of it. The purpose of the questioning was to find a way to contradict that thesis by developing a demonstration. If he manages to show a unitary relation in the chain of arguments, the conclusion was irrefutable. This is the birth of logic as a discipline: a progressive discursive refinement. Now picture practicing debates for centuries... as demonstrating a proposition meant to find a bridge (concept) to unify arguments, naturally that bridge needed to be more abstract than the proposition at stake. So the search for concepts, each time more abstract than the preceding ones, leaded to distinguish the most evanescent

abstractions ever thought by Occidental civilization. Philosophy appears relying on these cultural phenomena as an investigation of abstract concepts, mainly in the form of a written treatise [3]. That's the reason why Philosophy, as a discursive refinement, coordinate in the past the highest levels of abstraction of every other specific investigation. Many factors lead to the progressive separation of Natural Sciences, but for sure biographies such as Laplace's or Fourier's remind us of subjacent connections that may bring together apparently different things.

-[ES] That's true! I'm working in team that is developing a technology to convert the gradient of temperature in the ocean into energy. There are biologists, physicists, engineers, chemists and... some social science guys. Anyway, we are all trying to perform the best out of our specific knowledge, but there are times when problems call for an interaction. I mean, the other day the biologist asked for information about the pipes. It seems that an ecological disturbance can happen if you mix water of different densities.

-[P] Well, Oceanography is an excellent example of what is called nowadays inter-discipline. In 1902 the International Council for the Study of the Sea recognized that a sharp line should never be drawn between the disciplines that embraces all studies pertaining to the sea [4]. If we extend this criterion to the activity of the whole group of scientists in their particular subjects, we may see a wide-ranging and complex structures of interdependencies.[1] *But due to the accumulation of novel phenomena, a task of arrange and restate the results of different sciences needs to be done; in other words, "interdisciplinary inquiry requires understanding the manifold logical dependencies to which the regularities of different individual sciences give rise" [6]. Can you explain a bit why the biologist asked for that information to the engineers?*

-[ES] Well, it's part of the requirements of the project. I mean, each scientist contributes in their own area of expertise, but we are told about the caring of the environment... it's like the final filter. Now that I think about it, no matter how good your engineering calculus are done, a project may be completely rejected! Why is not enough the precision of the sciences?

-[P] We may need to recapitulate what led us to this situation. Since the XVIII century, the progress of sciences was meant to transform in moral progress of people: living better in community. The advances of science and technology were associated with reflection about freedom because of their practical consequences, such as improvement in hygienic conditions or material innovations. The question is how under these ideals are we now in a situation of environmental degradation and extended poverty all over the world? This is called the unintended consequences of development. For sure the precision that you talk about had led to reinforcement, through machines, of the normal performance of the human organism, but that has not result—paradoxically—in having control of the direction of the technical path we invented! That's why, nowadays, if we are not to repeat that historical process, we must take into consideration a wider connection of things involved. We are in the crossroad of figuring out if the means (the results of science and technology)

[1] Another useful distinction of terms is given by Spangenberg [5]: "I distinguish here between multidisciplinarity (several disciplines working parallel, with limited interaction, on a shared object of interest), interdisciplinarity (researchers from different disciplines working together in a way that their results can be integrated) and 'transdisciplinarity'. Transdisciplinarity is a demanding form of knowledge integration and it is dependent on reflectivity. It requires a non-idealized perception of the objects of research, a reflection on the specific limitations of the disciplinary construction of reality" (p. 277).

will determine the objectives, or objectives must be sketched first to apply just the necessary means. So precision in Natural Sciences is important but not in the same way or appliance to a social context.

2.1 The social object

This imaginary conversation is leading to our theme: how to analyze the social? Or even, what is "the social"? The mention of the historical passage about Fourier is relevant because it is the episode where a strange combination of scientific knowledge resulted in the consideration of social phenomena as something to do with behaviors, numbers, tendencies, and specific approaches: the study of people and their social relationships. Collections of data about populations (such as registers of property and use of land) are found since 1600. But the XIX century depicts a scenario where numbers about suicides per year were available and interpreted as an indicator of the quality of life. The second element was the debate if the notion of laws (such as Newton's) could be found in societies. Finally, a particular interpretation of probability (as a proposition about the stability of mass phenomena and the incorporation of the "law of errors" of observational astronomy) takes place in measuring the population characteristics [1]. For the purpose of this chapter, let us keep in mind two main ideas that remain about (a certain type of) analysis of society:

a. Governments obtain data from the population (e.g., criminality rates) in order to change the conditions that underlie certain social phenomena.

b. Knowing something is equivalent of measuring something.

The former is mainly a political claim that will be treated further. The latter certainly is a very strong tool, but we want to add some nuances to avoid the interpretation that without numbers there is no knowledge at all. Here are some broad considerations that distinguish society as an object of study [7]:

1. Society is immersed in historical contexts; it does not pertain to experimental systems of controlled conditions.

2. Partial relations in a society can be depicted by isolated magnitudes, but contexts of major complexity with interdependencies escape that treatment.

3. Investigation about society works within notions of comprehension, reference to values, and the co-elaboration of sense.

4. The social structure can be represented with variables such as demographic composition, social stratification, power positions, ways of production, and network of communications.

5. A theory of social action is meant to explain social processes in the dimension of historical processes.

6. An example of a type of reasoning (functional) about the social dimension could look like: If we examine a community, we recognize a social structure > the people (social units) are connected by a defined set of social relations > the continuity of the social structure is not destroyed by changes in the unities > the continuity of the structure is maintained by social life processes > these processes consist in the activities and interactions of individual human beings and

groups they form > a recurrent practice with a specific function (e.g., punishment) contributes to the maintenance of the social structure.

The main distinction to observe is the fact that the society can be an object (of investigation), just of different kind than the objects in Natural Sciences. Certainly the treatment of social mass phenomena in terms of quantifiable magnitudes—as Fourier's example—was the resultant of many other political needs of the time: statistical analysis of debt, annuities, life insurance rates, etc. This, all together, opened the possibility to interpret numbers about society as indicators and for information and planning in a broad way.

-[ES] So we are in an undesirable situation: scarcity of resources and humanitarian crisis. I kind of get what engineering can do for solving the problem. For example, OTEC technology could supply clean energy for coastal communities as well as fresh water. In that way the problem of energy supplying, water and reducing CO_2 emissions get solved... but I'm not sure exactly what a social analysis contribution would look like.

-[P] In Sociology, there are three traditions: one that tries to develop the categories of social action; the second is an analytic theory of action, which tries to clarify the structure of the activities meant to achieve a purpose (teleology); and the third, the theory of the speech acts, that departs from the processes of understanding between people [7]. All these knowledge is technical, because it develops a specialized language and, in one way or another, it produces information that may be traduced into technical recommendations. That is the level where empirical regularities—such as the ones of social statistics—can contribute, but there is a deeper level of analysis where an interpretation of the significance of social processes must be the outcome. Anyway, in the intersection of knowledge that we talk about earlier, it was in 1969 when the Special Committee on Problems of the Environment (SCOPE) took as its main investigation man's impact on natural ecosystems.[2] In 1971 the United Nations Educational, Scientific and Cultural Organization (UNESCO) devised its Man and the Biosphere (MAB) program where methods of data collection were meant to develop; 3 years later a convention was made to give a specific task to the contribution of Social Sciences; the main thesis was that man is simultaneously part of the environment and—with his awareness and capacity for deliberate action—responsible for its stewardship. Thus, man should not be seen merely as an "actor" on the environmental "stage." The "man in the biosphere" aspect should be stressed, and a systemic, holistic view taken on the interrelationships between man and other components of the biosphere [8].

So you stated the practical contribution of engineering in solving problems—which is true—but how could engineering help to solve the 500 environmental and social conflicts around the world concerning projects on electricity? [9].

-[ES] Well, but that is not part of engineering!

[2] The guiding question was: "In what qualitative ways and by how much quantitatively is the global environment of man being adversely affected by the technological revolution begun during the 19th century? The growth of population and the per capita capability of modification and exploitation of the environment are accelerating at such rapid rates that unforeseen and often unwanted side effects are continually arising. How serious are the consequences of these *for the future life of man on earth?*" [8]. SCOPE was the first environmental organization to focus attention on human influence.

-[P] You are right... but the conflicts remain as part of the reality of the world. This is where an interpretation of the significance of social processes take place: you may have the best OTEC prototype of the world but the people that is meant to receive the benefits of that material innovation is communicating something else, something that need to be attended before technical considerations. Here is an example. You may think that marine protected areas are intrinsically good because they help conservation and fisheries management. Somehow a protected area is a result of technical advises regarding environment, biodiversity, flows of energy, etc. But what about fishing and harvesting livelihoods? Studies have suggested [10] that such implementations may lead to negative experiences of the local communities such as results of marginal employment, monetary benefits from tourism for a select elite, reduce access to the area, and lack of development of social, cultural and financial assets. So there is a problematic relationship between the intention of solving or improving life conditions (as is the goal of engineering) and local communities, that eventually may lead in undermining the success of renewable energies.

-[ES] I get it. So Social Sciences can be used for avoiding a conflict?

-[P] Beyond that, because the intersection between renewable energy projects and the society should not be seen as a calculus to avoid conflicts, but as a scenario where new technologies fit human needs. And as you can guess, a human need is not merely access to energy but living in an integral context of wellbeing regarding health, prosperity, peace, and justice. In such an environment, technical innovations can be well received; without, hardly any innovation could substitute bonds within people, people with their institutions and people with nature.

The call for interdisciplinarity is a call for understanding better the complexity of today's world. Certainly, engineering plays a major role in the implementation of material improvements, but when negative experiences in the interaction technology-society start to proliferate, it is time to pose new questions. As society is the basis of our human reality, questions aiming at comprehension, values, and sense must be addressed. Social Sciences have theorized regarding all those topics. But as Plato once said, there is nothing more practical than theory. At any time in history, abstractions about society have found in daily life a correspondence and a feedback. We will see in the next section, how some of the general considerations about social analysis have landed in specific practical tools that concern the energy transition.

3. Social impact assessment tool

In this section, a brief critical history of the origin of the SIA and an interesting cycle of applications and reflections experienced by the practitioners of that tool during the 1970s and 1980s will be shown. Having in mind some critical perspectives and, indeed, actual debates regarding social impacts, the SIA will be presented in the framework of the International Association for Impact Assessment (IAIA) as a compilation of best practices. Finally, some specific tools for addressing socioeconomic variables will be presented.

3.1 A brief (critical) history

The history of Social Impact Assessment (SIA) begins in the 1970s, at the same time that new environmental legislation came into action via the National Environmental Policy Act (NEPA) in the United States of America. Measuring the effects of technological developments within parameters of quality of human life became relevant in the context where diagnosis of environmental degradation and complex processes of decision making were more and more intertwined. This was first appointed by the Environmental Impact Statement (EIS), concerning alterations in the biophysical environment and, only when social and economic impacts were "clearly interrelated," they were included in an EIS. This subject was ambiguous at the time, and became clearer as specific cases were treated. Here are two examples:

1. In 1983, the US Council of Environmental Quality required the Nuclear Regulatory Commission to prepare an EIS before restarting operations of a nuclear reactor at Three Mile Island. Some citizens argued that the psychological health of people of the area could be in jeopardy, but the US Supreme Court (*Metropolitan Edison Co. v. People Against Nuclear Energy*, 460 U.S. 766, 103 S. Ct. 1556 [1983]) did not find a close causal connection between the act of restarting the reactor and an increase in mental health problems. Although this was not solved in favor of a "social cause," it is the first antecedent of a SIA-related litigation.

2. In 1985, the Department of the Interior prepared an EIS on an affected area that involved the largest federal coal lease ever held. The Northern Cheyenne Tribe, inhabitants of that territory, sued the Department of the Interior for not including in their EIS likely social, cultural, and economic impacts of the project. They succeeded. A judge voided the sale of over 350 million tons of Federal coal (with a market price of 4 billion dollars) and rebuked the Department of the Interior for failing to deliver a meaningful analysis of impacts on certain groups of residents within the affected area of Federal coal with a market price of 4 billion dollars [11].

These precedents are relevant because it is in the context of interpretation and application of new legislation that the society and its different groups (private sector, indigenous groups, academia, government, etc.) make sense of their future interaction regarding construction projects. A unique, centralized, and unquestionable decision of industrial development—just for the sake of industrial development as an autonomous process of prosperity—would be no longer viable. Assuring that fundamental social structures are not torn apart while promoting material improvements is the goal of a Social Impact Assessment. So it is a contribution to broader policy considerations.

The discussion of what is a benefit or a damage in the social realm is a complex problem, a problem that may show a partial solution in a sense, but reveals negative consequences in another. For example, economic benefits of growth show results in solving some rural poverty and urban crises. A general indicator of employment may give signs of increase, but analyzing long-term effects can reveal, for example, concentration in less-skilled job categories, almost no decrease in local unemployment rate, and no influence on local youth to stay in their communities [11].

Precisely, in the history of SIA, there are cycles of understanding it as a tool. In the period 1970–1975, there was a strong tendency to develop guidelines, in which the American Sociological Association played an important role in using Social Sciences frameworks to clarify human-environment aspects in general terms. Later on, from 1975 to 1980, there was a bloom of application of SIA in a context of development of massive projects and energy development; procedural guidance was needed and all sorts of manuals were published with "all the information necessary to conduct a SIA" [11]. Finally, during 1981–1986 there was a reflection of what was known about social impacts, having empirical evidence accumulated during years. General problematic dichotomies were pointed out, such as: quantitative methods/qualitative methods, account of attitudes/non-attitudes, need for original data/use of publicly data, and use of many metrics/combining all impacts on a single metric. Some agreements of that epoch were:

1. It is inappropriate to limit a SIA to available data.

2. SIAs should not be limited to variables that are easily quantified and/or politically convenient to a certain group or person.

3. The summary to decision-makers should provide relevant information for judgment, even though it is expressed in many metrics; combining incommensurable metrics in favor of a simplistic "overall figure" should be avoided.

This brief and selective historical context has a purpose. Nowadays, there is a lot of material regarding procedural guidance on how to conduct a SIA. Reflections of the recent past should not be forgotten. The debate is still open between alternatives and the debate is still open within perspectives: help communities cope with (a certain type of) development or help developers learn about the livelihoods of non-urbanized communities. Are social scientists seen as internal proponents for community interests or as workers in favor of the developers' projects? Is it possible to identify a pro- or an anti-development bias in the SIA according to its legal structure? Should impacts be publicly revealed or maintained inside a private communicative sphere? Should social impacts be considered only during a project development or should they take into account long-term social problems (e.g., the erosion of a community's economic base) as well as changes in national policies? Should planning stay within the limits of *ceteris paribus* conditions or should other strategies be adopted for dealing with uncertainties?

3.2 The social impact assessment of the IAIA

The International Association of Impact Assessments (IAIA) in its *Guidance for assessing and managing the social impacts of projects* defines a Social Impact Assessment as "the processes of analyzing, monitoring and managing the intended and unintended social consequences, both positive and negative, of planned interventions (policies, programs, plans, projects) and any social change processes invoked by those interventions" [12]. It is emphasized that a SIA is not a product but a process of management that needs to be implemented at the same time of the conception and design of a project. Nowadays, social impacts are considered as modifications in people's way of life, culture, community, political system, environment, health and well-being, personal and property rights, and fears and aspirations.

Because all of these components belong to the intimate way of life of a community, a SIA must include participatory processes that, ideally, foment bonds of solidarity. Some of the key concepts relevant to a SIA stated by IAIA are:

3.2.1 Social License to Operate

As development nowadays is more than just meeting the regulatory requirements, leading corporations are encouraged to meet expectations of the local impacted communities as well as of governments and international ONG's. This term is mainly a language of business driving strategies to avoid reputational harm and, in the worst scenarios, strikes, protests, sabotages, and legal actions.

3.2.2 Free, Prior, and Informed Consent (FPIC)

Communication between a project developer and a local community must avoid coercion or manipulation of information (in favor of stakeholders). Consent must be sought before any technical intervention in the inhabited territory and complete information must be given to the parts explicitly stating the possible benefits and negative aspects of the project. The idea of self-determination of a community implies the possibility of saying no to a project. Reaching a consent means communities have a real option to say no or yes to a certain project.

3.2.3 Human rights-based approach

This measure is in line with the political philosophy of democratic regimes where human rights are at the core element of actions. Also, the demand for accountability and transparency, fostering empowerment, ensuring meaningful participation of right-holders, and guaranteeing non-discriminatory engagement, prioritizing vulnerable groups has become a clear mandate for any perspective of development.

3.2.4 Non-technical risks

Risks concerning managerial, legal, social, and political issues must not be minimized in contrast with physical, structural, and engineering risks.

3.2.5 Shared value

Considering not only conventional economic needs but societal needs is what gives meaning to the shared value. It is a way of re-considering the role of a company beyond profit and engaging with society benefits at the same time.

Within all these key concepts, IAIA guide suggests that there are four phases in a SIA, which are:

1. Understand the issues

2. Predict, analyze, and assess the likely impact pathways

3. Develop and implement strategies

4. Design and implement monitoring programs

For a deeper view on these subjects, IAIA guide is offered for free as a digital document. The considerations that concern this chapter are for introductory purposes in understanding some particularities of integrating social issues in engineering projects.

> *-[P] New regulations concerning social impacts, as you may see, try to encompass and preserve a delicate equilibrium. One that is similar to that of the physical environment, with balances and subtle ties that if not attended can lead to disastrous consequences.*

-[ES] I see... at least the same rigor and seriousness of a physical balance of flows is needed in the social realm. I know, I know, with its own concepts and procedures, but at the same level of importance. So I wonder how some of these procedures operate in coastal environments...

3.3 An outline for coastal contexts

Relationships between people and the ocean can shape the sense of place, personal identity, and a broad array of leisure, recreation, and work opportunities [13]. Increasingly, oceans are being used to meet resource needs. Marine energy development may play an important role in the redistribution of ownership rights in the marine environment changing the distribution of the current economic activities [14].

In this section, we identify the socioeconomic elements that may shape the public acceptance or perceptions related to marine renewable energy projects.

3.3.1 Public engagement

Support for renewable energy often stems from environmental concerns and ethical obligations related with fossil fuel generation and climate change consequences. Offshore renewable energy developments are often opposed due to concerns about the cumulative impacts of arrays of devices on landscapes and marine life. Also, the opposition may be based on a desire to keep the ocean free of industrial development to keep it intact [15]. However, public consultation can identify a timeline of the transformations of the territory. Yet, Social Sciences remain underutilized in energy policy, especially in a marine context [16]. Thus, creating alliances between stakeholders and the population could set the path to achieve agreements to promote sustainable development and design appropriate projects.

3.3.2 Disruptions of the landscape

Natural environments provide a wide variety of ecosystem services, which classify in (a) supporting services, (b) provision services, (c) regulating services, and (d) cultural services, this one is related to the spiritual attachment of the population or appreciation of the landscape. This service has a strong influence on the well-being of the communities, so the impact of visual disturbance is one of the aspects for opposing to a renewable energy project.

3.3.3 Changes in the economic activities of the area

Traditional maritime sectors are not always spatially compatible with the development of new maritime industries. In coastal zones, fishery is one of the main economic activities of the community. Possible restrictions and negative influence over the fish stocks are among the concerns in this area. Also, competition between different sectors for alternative uses of sea space can lead to sub-optimal economic development, which can cause conflicts between sectors and the rejection of the population and industries [17, 18]. While developers often meet only statutory requirements to acquire legal rights to occupy the sea space and exploit marine energy resources, they must also balance this with the rights and prerogatives of other sea users. There is little precedent regarding this subject due to minority of success in commercial developments. It would be useful to develop some guidance [19].

3.3.4 Energy supply

One of the targets of marine renewable energy is to establish a supply chain for the coastal communities to sustain the economic activities. The presence of Marine Renewable Energy (MRE) devices can address the current and future power needs of remote communities [20]. The expansion of the grid represents an increase in the cost of the projects, so developers are responsible for providing precise information of the range of the potential supply of the technologies and the percentage of the benefited population. Increased access to energy will improve the conditions of the communities such as: housing conditions, economical activities, services, etc. Energy access is variable and associated with local context. The access to this basic service may have positive repercussions and this could change the perception of the population regarding the modifications in the environment as a result of the establishment of an MRE extraction device.

3.3.5 Marine spatial planning (MSP)

The coastal zone is a particular area due to its inherent ecotone characteristics; coastal ecosystem has already experienced major changes due to human activities, triggering spatial conflict of uses and demands that are increasingly growing [21]. The issue manifests in the lack of mechanisms for monitoring, regulating, and legislating marine space. The marine spatial planning (MSP) is a multidisciplinary instrument for the promotion of sustainable development in the costal and marine space. It seeks an active society participation in the consultation phase with the purpose of identifying the socio-environmental problem [22]. Identifying the problems related to the different anthropogenic activities that are carried out in the coastal zone to assess its impact and determine if it is necessary to carry out an order to avoid conflicts between sectors is fundamental. Thus, the combination of ecological and human data in management instruments is particularly valuable in visualizing overlapping interest to achieve communication channels [20, 21, 23].

4. The Sustainable Development Goals

-[ES] Professor, you mentioned earlier how Fourier was committed to a sociopolitical project of his time, right? I guess that was his motivation for his technical work. Is there something similar today?

-[P] In a certain way there is. Have you heard about Sustainable Development? It is a political attempt, by all 193 member States of the United Nations, to impulse and prosperity, peace and partnership among people and promote caring for the planet. As geopolitical scenarios get more complicated, and the unintended consequences of development shows that poverty, inequality, violence and environmental degradation is the general case of the world, Sustainable Development rises as a pathway to prevent disastrous consequences. That's probably why Human Rights have climbed into such a determinant position even in front of development projects that involve great amounts of money, material resources and innovative technology... without a societal support, there are no pillars that could support development.

In 2015, the United Nations (UN) adopted the Agenda 2030 for Sustainable Development along with 17 Sustainable Development Goals (SDGs), which allow countries to concentrate their efforts toward common objectives. With regard to the energy sector, the UN included the Sustainable Development Goal 7 (SDG7), which aims at ensuring access to "affordable, reliable, sustainable and modern energy for all" [24]. Even though SDG7 stands as a goal in itself, it is of particular relevance because of its interrelation with the rest of the SDGs. In this sense, it is important to remember that energy is crucial for ensuring access to clean water and sanitation, health services, education, food, transportation, communications, and security, among others [22].

The SDGs are based on a holistic perspective of sustainable development that takes into account social, environmental, and economic concerns. In this sense, the SDGs recognize that the best way for addressing the most pressing challenges of the twenty-first century is through strategies that build economic growth and address the current social needs while at the same time they address climate change and focus on environmental protection [25]. With this perspective, the SDGs can be understood as goals that are interlinked and interrelated. This means that it is not possible to address one goal without also advancing in other goals.

With SDG7, the UN recognizes that energy is central for addressing every major challenge in the world today, from climate change to ending poverty, and indicates that access to energy for all is essential for sustainable development [26]. However, the UN points out that providing access to energy for all should be done through clean and sustainable sources of energy: "ensure access to affordable, reliable, sustainable and modern energy for all" [24], which can be expected to have an impact on the rest of the goals. In this sense, SDG7 can be considered as an enabling factor for addressing the rest of the SDGs [27].

To begin with, SDG7 aims to guarantee universal access to affordable, reliable, and modern energy services for all. This is crucial for vulnerable people around the world in order to be able to meet their basic needs and get access to basic services such as health care, education, information, adequate and safe housing, and access to drinking water and sanitation services (see **Figure 1**) [26]. Furthermore, energy is a fundamental input for daily activities such as lighting and electronic appliances whether they are for cooking, cleaning, leisure, or security [26].

One of the most pressing challenges in the world is to increase the access to clean cooking fuels and technologies. In 2017, only 61% of the global population had access to clean cooking fuels and technologies, which meant that the other 39%, almost 3 billion people, relied on inefficient and polluting cooking systems [24]. The use of wood, charcoal, and dung as fuel for cooking or for heating results in a constant interaction with smoke that includes a variety of pollutants and

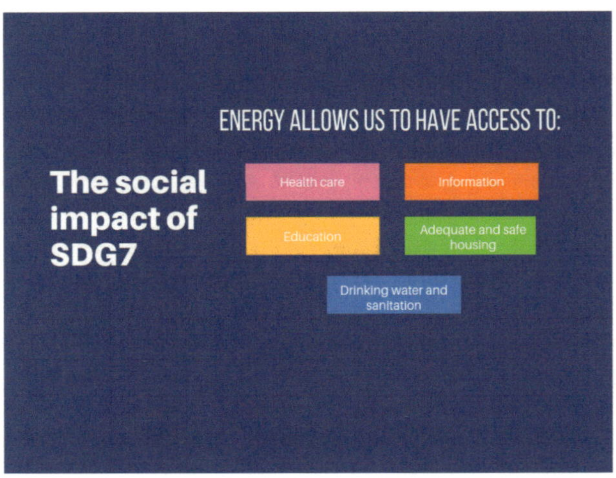

Figure 1.
Energy allows us to meet our basic needs and have access to basic services. Note: Graphic created by the authors.

Figure 2.
Social benefits of sustainable energy. Note: Graphic created by the authors.

translates into adverse health effects [26]. Likewise, increasing the access to clean and sustainable energy is particularly relevant for communities that are faced with extreme temperatures and need to use energy either in cooling or on heating systems [28].

The same can be said with regard to the transportation sector, which represents approximately 19% of the global use of energy and emits 23% of all the energy-related CO_2 emissions [28]. Whether it is by planes, buses, cars, trains, or ships, most of our modes of transportation still depend on fossil fuels and emit dangerous pollutants into the atmosphere that include greenhouse gases and particles that produce adverse health effects and have a negative environmental impact.

At the same time, guaranteeing universal access to affordable, reliable, and modern energy services could have contributions to reducing poverty, violence, and inequalities [26]. For example, an increment in the hours of access to electricity per day in communities with limited access to energy can have effects such as a higher number of people engaged in productive activities, longer hours of public lighting, which can increase safety at night, and reduction of the amount of hours dedicated to securing access to water, food, and heating/cooling, activities that are mostly carried out by women in the vulnerable sectors of the population (see **Figure 2**) [24, 26].

It is important to point out that the energy sector is the major contributor to climate change, representing approximately 60% of the total greenhouse gas emissions [29]. Therefore, transitioning toward clean and sustainable energy sources can have a significant impact in mitigating the adverse effects of climate change.

5. Conclusions

Renewable energies are in the middle of an ongoing debate about the means and the ends. The use of energy, especially since the Industrial Revolution, has been a fulcrum of an unparalleled material innovation that adds, not substitutes, the mechanical forces of a human body. Engineering has shown the amazing power of machines, which can be seen as the concrete expression of the most profound physical-mathematical abstractions articulated in ingenious, solid, and complex artifacts. Challenges concerning technical difficulties in the physical realm are incentives to the human mind to solve certain problems that derive from a feedback from nature: if a material does not resist or if a device malfunctions in particular conditions, we learn something from experience. We are to modify things in our procedures to test again our creations against wind, soil, and water.

But what about the social realm? Certainly, the same profundity in thoughts and abstractions is found in political theories, moral codes, and various philosophies. Material innovation is interrelated with social processes as much as biological phenomena in the ocean is interrelated with physico-chemical reactions. Ideas about human order (what should be the path to follow) encourage certain material arrangements and, in a cycle, material arrangements can modify and give birth to other ideas about the direction of progress. Our feedback from that social nature, in our statistics, is not uplifting. Lessons learned from the Industrial Revolution teach us that merely material innovation is not enough for solving deep societal problems in the world.

The debate between means and ends puts on the table a redefinition of the learning cycle of knowledge: proceed to let the social realm guide the technical innovations. To evaluate what "human needs" means requires not a unique paradigm, but flexible plans for different contexts of our multicultural world. That is one of the goals for the Social Impact Assessment: keeping possible a real scenario of communication and choice between a development proposal (that involves intervention in an inhabitant territory) and the people living in that particular place. It is a tool wide enough to interpret it as solely a new restriction and warning for profit-making under stricter rules, or a new way of making business that even benefits the profit-making as the world markets incorporate the environmental-social schemes, or as a political project that will try to show how it would look like if social parameters are the guiding line of the continuous industrial development, now, under regulations of environmental and human rights protection.

From these three perspectives, there are as many nuances as contexts in the world. General guides as the one offered by IAIA are very valuable sources that gather many lessons learned since the appearance of the SIA in the 1970s. This chapter intended to be a wider introduction for those themes, not pretending to substitute them but to offer some basic and interesting points to think about for an engineering student. The first section, presented mainly as a dialogue, meant to put into consideration what interdisciplinarity is about, at least in its attempt to bring together Engineering and Social Sciences. Afterward, we presented what can be understood, in general terms, by a social analysis. Broad considerations about the social object were stated just as examples to visualize some specific categories that contrast the vocabulary often used in Natural Sciences.

The brief historical preface of the SIA as a tool had two purposes: to reach some valuable reflections that came into surface during the 1970s and 1980s, when Environmental Impact Assessment included social and economic impacts as well. The other purpose was to show the value of historical research (part of Social Sciences) as people may be misguided by the apparent novelty of tendencies that, in fact, comes always from a historical process that has already clues of what is expected to be the result in the present.

From the general to the particular, we landed in a brief presentation of the SIA covering some fundamental perspectives and, later on, we gave a concrete example of a social assessment tool regarding marine environments. The last part concerning the SDG gives a closure to make sense why criteria about development are shifting from an apparent autonomous progress held by material innovations to the inter-linkage of objectives, both social and technical, to face extended worldwide poverty, inequities, and escalation of conflicts.

Social energy, as a metaphor of the relevance of our social structure, is making its way in the priorities of international agenda. For people acquainted with Social Sciences, these may be seen as something familiar and positive (yet, many adjustments are still needed [30]). People from other areas might perceive it as something strange and somehow imposing. Instead, it should be seen as an experiment, trying to take into consideration the feedback of history. It will be a long process of blending disciplines in which one may need wider perspectives to fully understand the phenomena at stake. Fourier or Laplace can give us clues of such a comprehension, and models for the engineering students of the future.

Acknowledgements

This work was encouraged by the Mexican Center of Innovation in Ocean Energy (CEMIE-O), which has embraced, for a couple of years now, an annual cycle of international conferences called Sustainable Energy: Impact and Social Responsibility, from where many of the reflections on social issues in this text came. We are grateful to Dr. Rodolfo Silva Casarín, Technical Director of CEMIE-O, and Dr. Miguel Ángel Alatorre Mendieta, Leader of the Thermal Gradient Team in CEMIE-O.

Author details

Alonso Pérez Pérez[1*], Mauricio Latapí Agudelo[2] and Graciela Rivera Camacho[1]

1 Mexican Center of Innovation in Ocean Energy (CEMIE-O), Mexico City, Mexico

2 University of Iceland, Reykjavík, Iceland

*Address all correspondence to: alonsopp@ciencias.unam.mx

References

[1] Hacking I. The Taming of Chance. USA: Cambridge University Press; 1990. 264p. ISBN: 9780521388849

[2] Miller T. Constructing sustainability science: Emerging perspectives and research trajectories. Sustainability Science. 2013;**8**:279-293. DOI: 10.1007/s11625-012-0180-6

[3] Colli G. La Nascita Della Filosofía. Milan: Adelphi; 1975. p. 116

[4] Sverdrup H, Johnson M, Fleming R. The Oceans. Their Physics, Chemistry, and General Biology. New York: Prentice-Hall, Inc; 1946. 1087p

[5] Spangenberg J. Sustainability science: A review, an analysis and some empirical lessons. Environmental Conservation. 2011;**38**(3):275-287. DOI: 10.1017/S0376892911000270

[6] Pieterinen A. Interdisciplinarity and Peirce's classification of the sciences: A centennial reassessment. Perspectives on Science. 2006;**14**:127-152. DOI: 10.1162/posc.2006.14.2.127

[7] Habermas J. On the Logic of the Social Sciences. USA: MIT University Press; 1990. p. 238

[8] Haas E, Williams M, Babai D. Scientists and World Order. The Uses of Technical Knowledge in International Organizations. USA: University of California Press; 1977. p. 368. ISBN: 0-520-03341-8

[9] Environmental Justice Atlas [Internet]. 2019. Available from: https://ejatlas.org/commodity/electricity [Accessed: 10 August 2019]

[10] Bennett N, Dearden P. Why local people do not support conservation: Community perceptions of marine protected area livelihood impacts, governance and management in Thailand. Marine Policy. 2014;**44**:107-116. DOI: 10.1016/j.marpol.2013.08.017

[11] Freudenburg W. Social impact assessment. Annual Review of Sociology. 1986;**12**:451-478

[12] Vanclay F, Esteves A, Aucamp I, Frank D. Social Impact Assessment: Guidance for Assessing and Managing the Social Impacts of Projects. Fargo ND: International Association for Impact Assessment; 2015

[13] McKinley E, Acott T, Stojanovic T. Socio-cultural dimensions of marine spatial planning. In: Zaucha J, Gee K, editors. Maritime Spatial Planning. Cham: Palgrave Macmillan; 2019

[14] Kerr S, Watts L, Colton J, Conway F, Hull A, Johnson K, et al. Establishing an agenda for social studies research in marine renewable energy. Energy Policy. 2014;**67**:694-702. DOI: 10.1016/j.enpol.2013.11.063

[15] Jung N, Moula ME, Fang T, Hamdy M, Lahdelma R. Social acceptance of renewable energy technologies for building in the Helsinki metropolitan area of Finland. Cogent Environmental Science. 2016;**99**:813-824. DOI: 10.1016/j.renene.2016.07.006

[16] Magagna D, Uihlein A. Ocean energy development in Europe: Current status and future perspectives. International Journal of Marine Energy. 2015;**11**:84-104. DOI: 10.1016/j.ijome.2015.05.001

[17] Sannino G, Pisacane G. Ocean Energy Exploitation in Italy: Ongoing R&D Activities. Italy: ENEA; 2007. 54p

[18] Rudolph D, Haggett C, Aitken M. Community benefits from offshore renewables: the relationship between different understanding of impact, community, and benefit.

Environment and Planning C: Politics and Space. 2017;**36**(1):92-117. DOI: 10.1177/2399654417699206

[19] Kerr S, Watts L, Brennan R, Howell R, Graziano M, O'Hagan AM, et al. Shaping blue growth: Social sciences at the nexus between marine renewables and energy policy. In: Foulds C, Robison R, editors. Advancing Energy Policy, Lessons on the Integration of Social Sciences and Humanities. Cambridge: Palgrave Macmillan; 2018. pp. 31-46. DOI: 10.1007/978-3-319-99097-2_3

[20] Commonwealth secretariat. Marine Renewable Energy, Commonwealth Blue economy series No. 4. United Kingdom: Malborough house; 2016. p. 55

[21] Azzellino A, Conley D, Vicinanza D, Kofoed JP. Marine renewable energies: perspectives and implications for marine ecosystems. The Scientific World Journal. 2013;**2013**:1-3. DOI: 10.1155/2013/547563

[22] European commission. Energy Sector and the Implementation of the Maritime Spatial Planning Directive. Belgium: European Union; 2015. 20p

[23] Ban NC, Bodtker KM, Nicolson D, Robb CK, Royle K, Short C. Setting the stage for marine spatial planning: Ecological and social data collation and analyses in Canada's Pacific waters. Marine Policy. 2013;**39**:11-20. DOI: 10.1016/j.marpol.2012.10.017

[24] United Nations. Sustainable Development Goal 7 United States of America: United Nations [Internet]. 2017. Available from: https://sustainabledevelopment.un.org/sdg7 [Accessed: 05 August 2019]

[25] United Nations. The Sustainable Development Agenda: United Nations [Internet]. 2018. Available from: https://www.un.org/sustainabledevelopment/development-agenda/ [Accessed: 15 August 2019]

[26] United Nations. SDG7: Ensure Access to Affordable, Reliable, Sustainable and Modern Energy for All: United Nations [Internet]. Available from: https://www.unescap.org/our-work/energy/energy-sustainable-development/about [Accessed: 01 August 2019]

[27] Alloisio I. SDG 7 as an enabling factor for sustainable development [Power Point Slides]. 2018. In press

[28] Sustainable Energy for All. Understanding SDG7 [Internet]. Available from: https://www.seforall.org/understanding-sdg-7 [Accessed: 10 August 2019]

[29] United Nations. Sustainable Development Goal 7: United Nations [Internet]. 2019. Available from: https://www.un.org/sustainabledevelopment/energy/ [Accessed: 16 August 2019]

[30] Palsson G, Szerszynski B, Sörlin S, Marks J, Avril B, Crumley C, et al. Reconceptualizing the 'Anthropos' in the Anthropocene: Integrating the social sciences and humanities in global environmental change research. Environmental Science & Policy. 2013;**28**:3-13. DOI: 10.1016/j.envsci.2012.11.004

Section 2

Technical and Theoretical

Chapter 3

Current Development and Prospect of Turbine in OTEC

Weimin Liu, Yunzheng Ge, Lei Liu and Yun Chen

Abstract

This chapter mainly introduces the development and prospect of turbines utilized in ocean thermal energy conversion (OTEC), including brief introduction, aerodynamic design, mechanical and electric control system, problems, and prospect of the turbine in OTEC. At the beginning, the first section mainly introduces compositions and types of turbine in OTEC systems, different working fluids in the turbine, status of OTEC turbine currently in the world, and so on. After that, the aerodynamic design of turbine has the greatest impact on the efficiency of the turbine and is the most important process during the turbine development. Therefore, the second section focuses on turbine aerodynamic design. Furthermore, the third section mainly introduces the mechanical system and the electric control system of the turbine. Finally, the problems in the turbine currently are listed, and the prospects for the development of the turbine in OTEC are discussed in the fourth section.

Keywords: OTEC, radial-inflow turbine, aerodynamic design, mechanical system, electric control system

1. Introduction

This section mainly introduces compositions and types of turbine in ocean thermal energy conversion (OTEC) systems, different working fluids in turbine, current status of OTEC turbine in the world, and so on. In OTEC system, a turbine rotates at a high speed driven by a heated gas and then drives the generator to generate electricity. So turbine is the key equipment that converts ocean thermal energy into mechanical energy and finally output electricity. Moreover, the output power and thermal efficiency of OTEC system change with that of turbine, so the performance of turbine may directly affect the whole performance of OTEC system.

Since the temperature of working fluid flowing through the turbine such as ammonia, water, and so on is low and the enthalpy drop is small, OTEC power generation system is different from other high-temperature power cycles like steam Rankine cycle or gas turbine cycle. The inefficiency of overall power generation system about 3–5% requires that the turbine must have high efficiency. Since the efficiency of OTEC system which adopts an open cycle is low, a closed cycle is usually used to obtain more electricity. And in closed cycle, internal working environment is relatively good for turbine, and design difficulty is relatively low. In addition, since organic working fluid or ammonia is used as working fluid, turbine requires high performance of sealing, and sealing material should not be affected by working fluid. Furthermore, special lubricating oils suitable for organic working

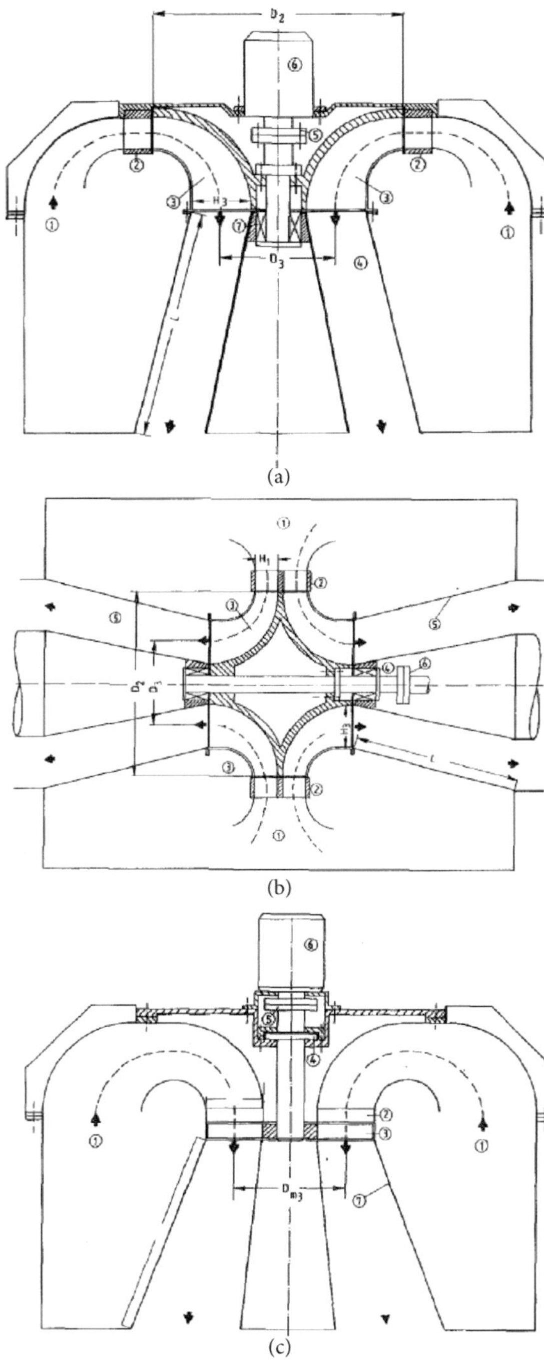

Figure 1.
Different types of turbines (a) single-inflow radial steam turbine: 1 = inlet from evaporator, 2 = stator, 3 = rotor, 4 = exit diffuser, 5 = coupling, 6 = generator, 7 = bearing, (b) double-inflow radial steam turbine: 1 = inlet from evaporator, 2 = stator, 3 = rotor, 4 = bearing, 5 = exit diffuser, 6 = coupling, (c) single-stage axial steam turbine: 1 = inlet from evaporator, 2 = stator, 3 = rotor, 4 = bearing, 5 = coupling, 6 = generator, 7 = exit diffuser.

fluids or ammonia should be selected. For large-scale systems, special lubricating oil separation devices or lubricating oil systems should also be used.

The efficiency of axial turbine, which was once used in open cycle of OTEC system, is lower than that of radial type. For example, the efficiency of single-inlet and double-inlet radial turbine in the Hawaii Natural Energy Institute (HNEI) is increased by about 10% compared with that of axial turbine in Ref. [1], which is shown in **Figure 1**. Therefore, it is recommended to adopt radial turbine in open cycle. Similarly, radial turbine is often used in closed cycle of OTEC power generation system, considering the requirements of improving turbine efficiency and impeller strength.

In closed cycle of OTEC system, different working fluids are used depending on different systems. For Rankine cycle, organic working fluids are usually used as working fluid, such as R600a, R245fa, R152a, R22, and R134a, in Ref. [2, 3]; in Karina cycle and Uehara cycle, ammonia-water mixture is adopted as working fluid in Ref. [4, 5]. Therefore, in OTEC systems adopting different thermodynamic cycles, most of the working fluids in turbine are ammonia or other organic working fluids. The main difference is that for system using ammonia-water mixture, working fluids in turbine are almost pure ammonia, so turbine design can be carried out according to that using pure ammonia as working fluid.

The turbine used in 30 kW OTEC system of Saga University in Japan is a two-stage radial turbine with intermediate extraction. For 100 kW OTEC power generation unit built in Maja area of Kume Island, radial turbine is adopted as well as in 100 kW OTEC power plant built by Makai Ocean Engineering in Hawaii. In addition, radial turbines are also used as key equipment to drive generators in OTEC test systems of Korea [6] and China [7]. In view of this, this chapter mainly studies the status, design, and prospect of radial turbine in a closed OTEC system.

2. Aerodynamic design of turbine in OTEC

This section focuses on turbine aerodynamic design in OTEC. In offshore OTEC power generation system which adopts closed cycle, the temperature difference between surface warm water (25°C) and deep cold water (5°C) is only about 20°C, which results in a small temperature difference of working fluid at the inlet and outlet of turbine. Even if ammonia with high expansion ratio is used, the pressure ratio between inlet and outlet of turbine is still very low. At the same time, due to low efficiency of OTEC thermodynamic cycle, the efficiency of turbine restricts efficiency improvement in OTEC power plant. So in the turbine design process, the influence of aerodynamic design on turbine efficiency improvement is the key content of turbine research.

2.1 Principle and characteristics of turbine in OTEC

The radial turbine is a high-speed and rotating machine, which can convert thermal energy of high-temperature and high-pressure working fluid into mechanical energy. The working fluid flows from turbine inlet to volute at a certain initial speed, and then the airflow is gradually distributed evenly during the flow process, and furthermore the gas expands and accelerates to obtain a higher speed in the guide nozzle. The guide is mainly a nozzle ring composed of many fixed stator blades, namely, radial guide nozzle. The velocity of gas at nozzle outlet increases, and the pressure and temperature decrease. And then the gas flows into the high-speed rotating impeller at a relative speed. The gas continues to expand in impeller, and the pressure of working fluid at impeller outlet is reduced. Finally, working fluid is discharged to the subsequent process through outlet diffuser.

The radial turbine in OTEC system has the following characteristics:

1. Both residual velocity loss and flow loss inside the turbine are relatively small, so it has higher internal efficiency.

2. Working fluid flow can be adjusted by adjustable nozzle, so that the turbine has a wider operating range and enhances the load adaptability and operation stability of the whole system.

3. The efficiency will not be greatly affected by the inaccuracy of blade geometry and surface smoothness. In addition, the turbine has a lighter weight, a simple structure, and fewer parts, which can greatly reduce manufacture cost.

4. The power of radial turbine is limited to some extent by structural factors due to its large radial dimension and high speed.

Based on the above features, radial turbine is especially suitable for small flow rate and low pressure ratio, which is exactly the requirement for turbine of OTEC.

2.2 One-dimensional thermodynamic design and optimization of radial turbine in OTEC

The aerodynamic design of radial turbine mainly includes one-dimensional (1-D) thermodynamic design and three-dimensional (3-D) aerodynamic design of pneumatic parts including volute, nozzle, impeller, and diffuser as shown in **Figure 2**. Many works like [8–11] have introduced the design theory of radial turbine in detail; the readers can learn from them according to research situation. And this chapter only lists a design method for turbine, for example. 1-D thermodynamic design methods for OTEC turbine have been discussed in many literatures, including mean-line model, as shown in [12–15].

The flow of working fluid at the inlet and outlet of impeller can be expressed by velocity triangle shown in **Figure 3**. Absolute velocity $\boldsymbol{c}_\mathrm{j}$ is the vector sum of circumferential velocity $\boldsymbol{u}_\mathrm{j}$ and relative velocity $\boldsymbol{\omega}_\mathrm{j}$ at inlet and outlet of impeller. The working fluid flows into impeller in direction of absolute flow angle α_1 at the inlet and flows out of impeller in direction of relative flow angle β_2 at outlet. At impeller inlet, relative flow angle β_1 is the angle between relative flow direction and u_1, while the angle between absolute flow direction and u_2 at the outlet of impeller is the absolute outlet flow angle α_2.

Firstly, the main 1-D thermodynamic and geometric parameters of radial turbine are calculated based on the given initial data and some other important empirical parameters. Generally speaking, the most important design parameters of turbine include the wheel diameter ratio $\overline{D}_2$, degree of reaction Ω, and velocity ratio $\overline{u}_1$.

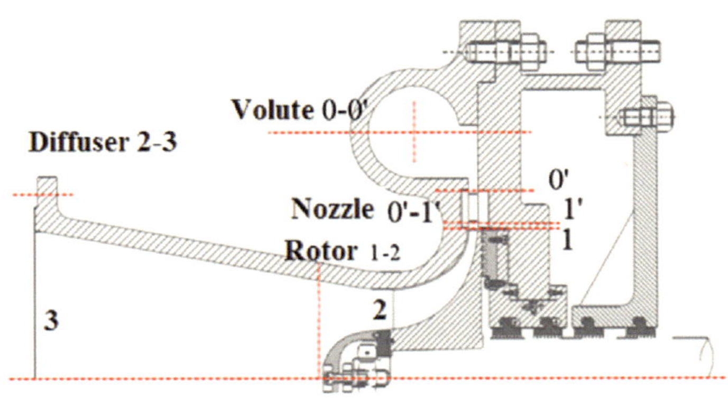

Figure 2.
Schematic diagram of turbine pneumatic structure.

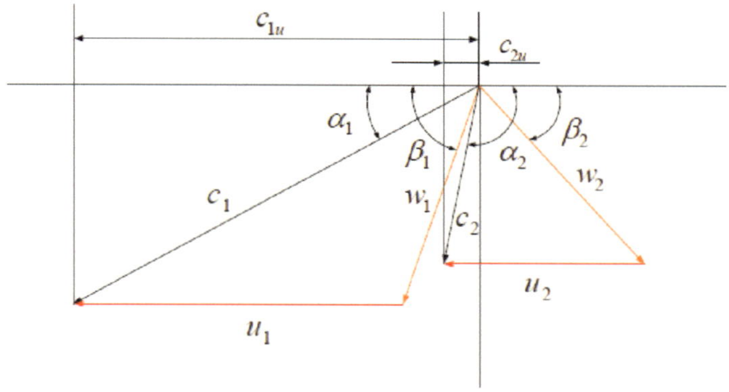

Figure 3.
Velocity triangle of radial turbine impeller.

The wheel diameter ratio refers to the ratio of impeller outlet diameter to inlet diameter, and the wheel diameter ratio determines the turbine working ability, which is generally selected between 0.35 and 0.55. Degree of reaction refers to the ratio of the isentropic enthalpy drop in impeller to the total isentropic enthalpy drop in turbine, which represents the distribution of energy when gas expands in nozzle and impeller. A larger degree of reaction means a faster velocity of the airflow discharged from impeller and more residual velocity loss. When the degree of reaction is too small, deceleration motion will occur. For turbine, it is generally selected between 0.3 and 0.5. The velocity ratio is the ratio of circular velocity at impeller inlet to the ideal velocity under isentropic condition, which reflects the influence of rotational speed on turbine, generally selected between 0.5 and 0.8.

There are many ways to determine the thermodynamic parameters of turbine, such as trial method, optimal velocity ratio method, and screening method. The trial method refers to the calculating wheel efficiency using selected parameters such as degree of reaction, wheel diameter ratio, velocity ratio, inlet nozzle flow angle, and impeller outlet flow angle in order to determine the optimal design. This method has great blindness and requires a lot of work. The optimal velocity ratio method determines the main design parameters such as the velocity ratio and the degree of reaction by interpretation method, such as "zero tail vortex" analysis method and specific velocity analysis method. The screening method is used to analyze the effect of the main parameters such as pressure ratio, Mach number, angle of attack, and allowable stress of wheel in determining turbine velocity ratio and degree in detail. This method has high efficiency and can prevent large changes.

In this chapter, the design of a turbine using ammonia as working fluid for 7.5 kW OTEC is shown as an example. According to the calculation result of cycle, the given thermodynamic design parameters of turbine are shown in **Table 1**.

In order to obtain 1-D design parameters, the important parameters to be determined include wheel diameter ratio $\overline{D}_2$, degree of reaction Ω, velocity ratio $\overline{u}_1$, velocity coefficient, and blade angle. Velocity coefficient includes those of impeller φ and nozzle ψ. Blade angle includes impeller inlet mounting angle α_1 and nozzle outlet mounting angle β_2 in [16]. The values of these seven basic parameters are selected by the above various methods in proper range. Finally, the range and values of the seven basic parameters selected are listed in **Table 2**. Based on design parameters, thermodynamic parameters in turbine are calculated and 1-D thermodynamic design is completed. The result is shown in **Table 3**.

In the process of 1-D design above, the parameters are determined by using traditional methods. There are, however, limitations in the methods, such as time-

Parameters	Units	Values
Gas constant	N.m/(kg.K)	488.2
Isentropic exponent		1.447
Relative molecular mass		17
Mass flow rate	kg/s	0.388
Inlet pressure	MPa	0.9
Inlet temperature	K	298
Outlet pressure	MPa	0.75

Table 1.
The design parameters of turbine.

Parameters	Units	Ranges	Values
Ω		0.35–0.55	0.48
$\overline{D}_2$		0.3–0.5	0.38
$\overline{u}_1$		0.65–0.7	0.656
φ		0.92–0.98	0.96
ψ		0.75–0.9	0.84
α_1	°	12–30	15
β_2	°	20–45	35

Table 2.
The range and values of the basic parameters.

consuming, blindness, and even unreasonable. Therefore based on design parameters obtained through the above methods, an optimization of the relevant parameters is carried out in order to obtain better design parameters. The optimization method mainly determines value ranges of the seven design parameters and establishes the following optimization model as Eq. (1).

$$\eta = 2\overline{u}_1\left(\varphi\cos\alpha_1\sqrt{1\text{-}\Omega\text{-}\overline{D}_2^{\,2}\overline{u}_1} + \overline{D}_2\psi\cos\beta_2\sqrt{\Omega + \varphi^2(1\text{-}\Omega) + \overline{D}_2^{\,2}\overline{u}_1^{\,2}\text{-}2\overline{u}_1\varphi\cos\alpha_1\sqrt{1\text{-}\Omega}}\right) \quad (1)$$

For the above model, an optimization algorithm is needed to obtain the proper solution. **Table 4** lists the results of 1-D thermodynamic optimization using a genetic algorithm program. Compared with the results in **Table 3**, the size of nozzle and impeller has changed slightly, and the airflow angle has changed correspondingly, which results in the reduction of output and a little increase in isentropic efficiency. Although these parameters change a little, improvement of isentropic efficiency shows that the optimal design has obtained a better result.

2.3 Three-dimensional design, simulation, and optimization of radial turbine in OTEC

Aerodynamic design has the greatest impact on the efficiency of turbine and is the most important part in turbine design. Therefore, this part focuses on turbine aerodynamic design and optimization.

Parameters	Units	Results
Impeller speed	r/min	21,000
Nozzle outlet velocity	m/s	155
Nozzle outlet pressure	MPa	0.802
Nozzle outlet temperature	K	291.08
Impeller inlet height	mm	5.8
Impeller inlet absolute flow angle	°	14
Impeller inlet relative flow angle	°	86
Impeller inlet relative velocity	m/s	37
Impeller inlet diameter	mm	120
Wheel peripheral speed	m/s	156
Impeller outlet absolute flow angle	°	101.22
Impeller outlet absolute speed	m/s	49.16
Impeller outlet relative flow angle	°	35
Impeller outlet external diameter	mm	56.2
Impeller outlet inner diameter	mm	31.7
Power generation	kW	7.8
Efficiency of the wheel periphery	—	0.864

Table 3.
The result of 1-D thermodynamic design.

Parameters	Units	Results
Impeller speed	r/min	21,000
Nozzle outlet velocity	m/s	160.5
Nozzle outlet pressure	MPa	0.785
Nozzle outlet temperature	K	290
Impeller inlet height	mm	4.398
Impeller inlet absolute flow angle	°	16
Impeller inlet relative flow angle	°	90
Impeller inlet relative velocity	m/s	40.9
Impeller inlet diameter	mm	126.8
Wheel peripheral speed	m/s	159.4
Impeller outlet absolute flow angle	°	99.364
Impeller outlet absolute speed	m/s	49.16
Impeller outlet relative flow angle	°	35.745
Impeller outlet external diameter	mm	57
Impeller outlet inner diameter	mm	28.8
Power generation	kW	7.65
Isentropic efficiency	—	0.875

Table 4.
The results of 1-D thermodynamic optimization.

2.3.1 Three-dimensional aerodynamic design of radial turbine in OTEC

Based on 1-D parameters obtained in the abovementioned optimization process, 3-D structural design needs to be carried out for aerodynamic components of turbine. The role of volute is to distribute airflow evenly, so the cross-sectional area of volute decreases continuously, which could be realized by using spiral type. In this case, asymmetrical pear-shaped cross section is adopted for volute design in consideration of good guidance of airflow. Compared with other cross-sectional volutes, this type has the characteristic of good flow performance although the entire volute has a larger size. **Figure 4** shows the designed 3-D model of the asymmetrical pear-shaped volute.

Nozzles of turbine are mainly used to make impeller capture the maximum momentum as effectively as possible, so nozzles always need a certain inclination angle. Nozzles have many forms, such as airfoil, straight plate, wedge, island, and so on, but some of these blades have large flow loss or poor aerodynamic effect, which cannot meet the requirements of high-efficiency turbine. In order to achieve a high-efficiency aerodynamic performance of turbine, the NASA-designed Rotor series andthe Soviet Union's TC-P series are selected as the original design profiles to obtain the required nozzles optimized. In this case, the TC-2P blade of the Soviet Union is selected to shape the nozzle, and the result is shown in **Figure 5**.

Meridian parameters are the basis of 3-D impeller design. On the basis of 1-D parameters, meridian parameters of the impeller, as shown in **Figure 6**, are calculated firstly as listed in **Table 5**. There are many references for 3-D design of impeller, and kinds of corresponding software are designed using internal algorithms such as ANSYS BladeGen and NREC. In this case, the 3-D design of impeller is carried out in two parts: the working wheel and the induce wheel. Generally, for strength, radial straight blades are used as working wheels. However, nonradial blades are preferred for flow, as shown in the literature [13, 15]. There are requirements for induce wheel such as good aerodynamic performance, geometric shape to meet the flow, better strength characteristic, and feasibility of production. For a better aerodynamic performance, parabolic surface is mainly used for blade modeling of induce wheel. Because the non-developable ruled parabolic of cylindrical base surface has good properties in aerodynamics, strength, process ability, and versatility, it can meet the basic requirements of the induce blade profile. In this case, the induce blade is modeled by a non-developable linear parabolic surface with a cylindrical base. The final designed impeller is shown in **Figure 7**.

For the design of diffuser, it is mainly considered that the gas discharged from turbine impeller still has a high speed, which will cause a large friction loss when

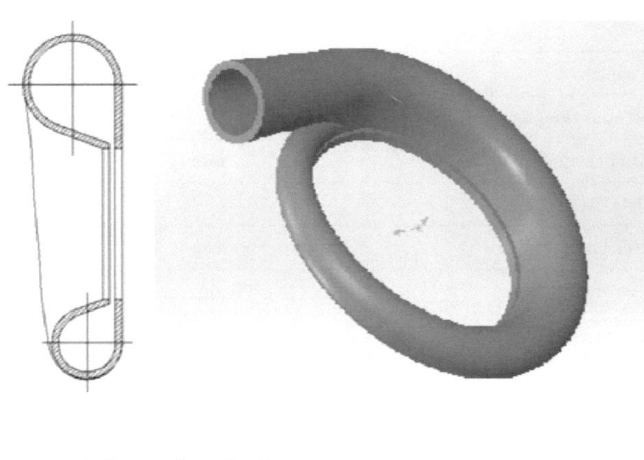

Figure 4.
The 3-D design of asymmetrical pear-shaped volute.

Figure 5.
The 3-D design of nozzle in turbine.

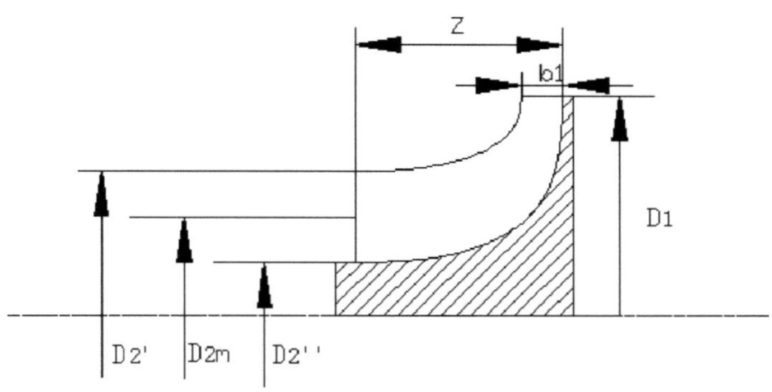

Figure 6.
The meridian plane of impeller.

Parameters	Units	Values
D_1	mm	117
D_2'	mm	57
D_2"	mm	28.8
D_{2m}	°	42.9
Z	mm	27
b_1	mm	4.43
Number of blades	—	12

Table 5.
The parameters of impeller meridian.

discharged into pipeline. Therefore, the function of diffuser is mainly to reduce the airflow velocity further in order to reduce airflow loss. Diffuser usually adopts a relatively simple conical cylinder with a certain expansion angle, and the

Figure 7.
The 3-D design of impeller.

cross-sectional area is calculated similarly to that of nozzle. **Figure 8** shows the designed diffuser, and the final 3-D design of turbine is shown in **Figure 9**.

2.3.2 Numerical simulation and performance analysis of radial turbine in OTEC

When turbine aerodynamic design in OTEC is completed, numerical simulation is used to study the aerodynamic performance of turbine. The aerodynamic part of turbine needs to be meshed before numerical simulation of turbine aerodynamic performance, and meshing can be carried out in many kinds of commercial software, including ICEM, TurboGrid, etc. The meshing of turbine nozzle and impeller can be carried out by Turbogrid, which can divide the structured grid with high grid quality and a short time. In this case, Turbogrid is used for meshing of nozzle and impeller, by dividing the structured grid with high grid quality and a short time. On the other hand, meshing of volute and diffuser is carried out in ICEM, as it can be used for regular and irregular geometric structures. **Figure 10** shows the grid of turbine.

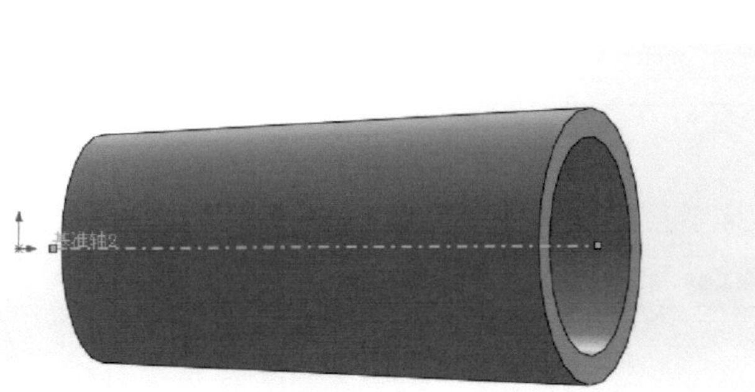

Figure 8.
The 3-D design of diffuser.

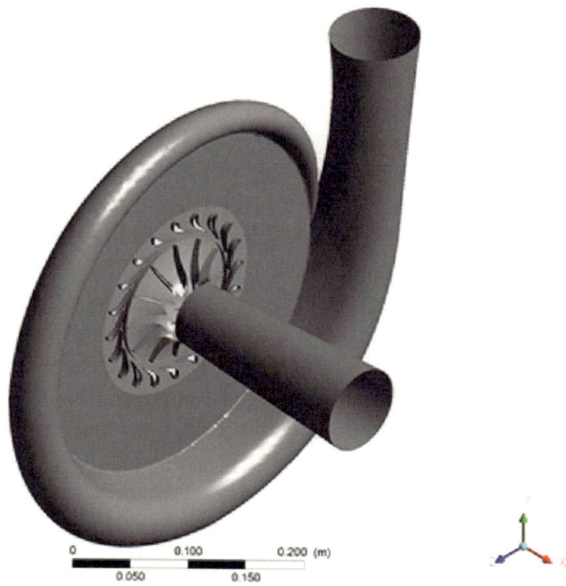

Figure 9.
The 3-D model of turbine.

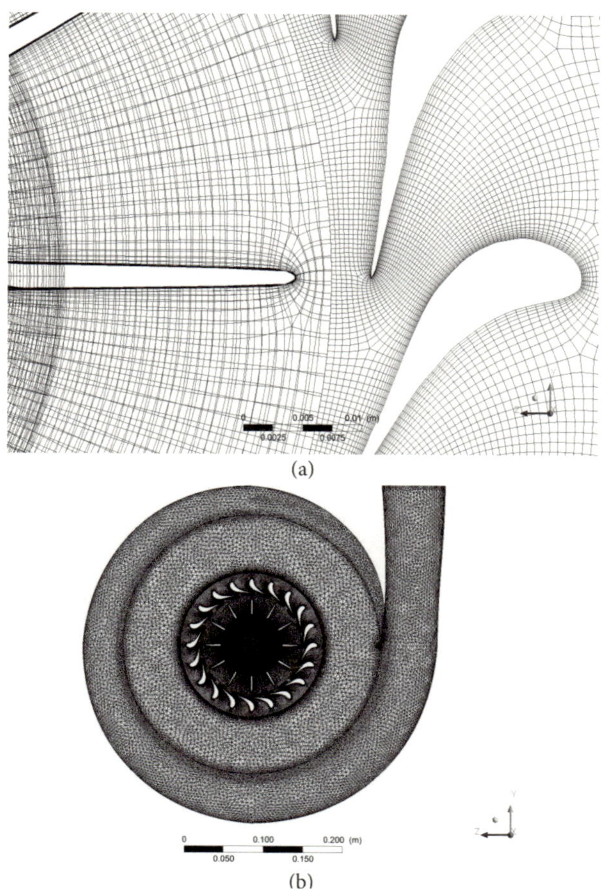

Figure 10.
Grid of radial turbine (a) hexahedral mesh of nozzle and impeller, (b) whole grid of turbine.

After overall meshing of turbine is completed, numerical simulation of turbine needs to be carried by commercial software, such as ANSYS CFX, ANSYS Fluent, and NUMECA. And these kinds of software are basically consistent in the simulation process, and internal algorithms include flow control equation, turbulence model, and solver. In this case, ANSYS CFX is used to simulate the turbine flow, including setting turbulence model, boundary conditions, simulation accuracy, and boundary condition initialization. The simulation method section is not repeated here, since it has been described in various kinds of simulation software.

In order to determine the accuracy of 1-D design, the performance of turbine needs to be simulated and analyzed, and then compared the results with that of 1-D design, as shown in **Table 6**. From the results, it can be seen that the simulation values of the main design parameters are basically the same as those of the design parameters except that the error of nozzle exit velocity is about 5%. Although isentropic efficiency and power are smaller than design values, the error is about 1%, which meets the design requirements.

Thereafter, the flow state of turbine is analyzed through streamline diagram, pressure contour, velocity contour, and Ma number contour. **Figure 11** shows a streamline diagram of turbine, and **Figure 12** visualizes pressure and velocity contour of nozzle and impeller passage section at 50% blade spanwise. As seen from the figures, the overall pressure gradient direction is roughly along the flow direction, the streamline flows uniformly along the nozzle and impeller channel direction, and no obvious back pressure gradient appears in the impeller. On the whole, although flow loss exists at the suction surface of blade, the efficiency of the turbine is higher,

Parameters	Design results	CFD results	Errors (%)
Nozzle outlet velocity ($m{\cdot}s^{-1}$)	160.5	152.5	4.98
Wheel peripheral velocity ($m{\cdot}s^{-1}$)	159.4	156.3	1.94
Mass flow rate ($kg{\cdot}s^{-1}$)	0.388	0.39	0.51
Isentropic efficiency ()	0.875	0.865	1.14
Power (kW)	7.65	7.54	1.44

Table 6.
Comparison of simulation results with design results.

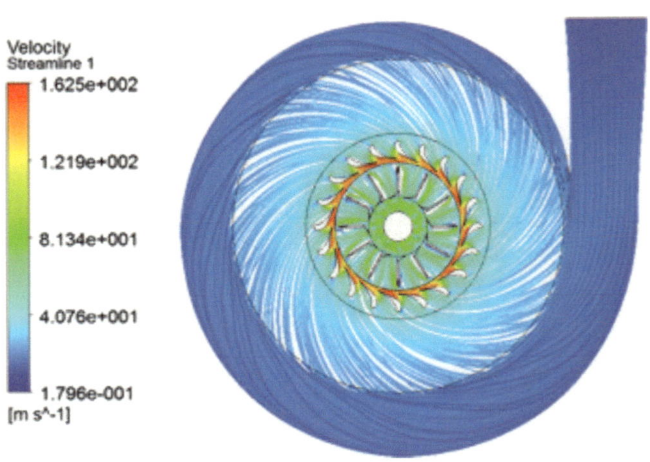

Figure 11.
Turbine overall streamline without including diffuser (a) pressure contour, (b) velocity streamline diagram.

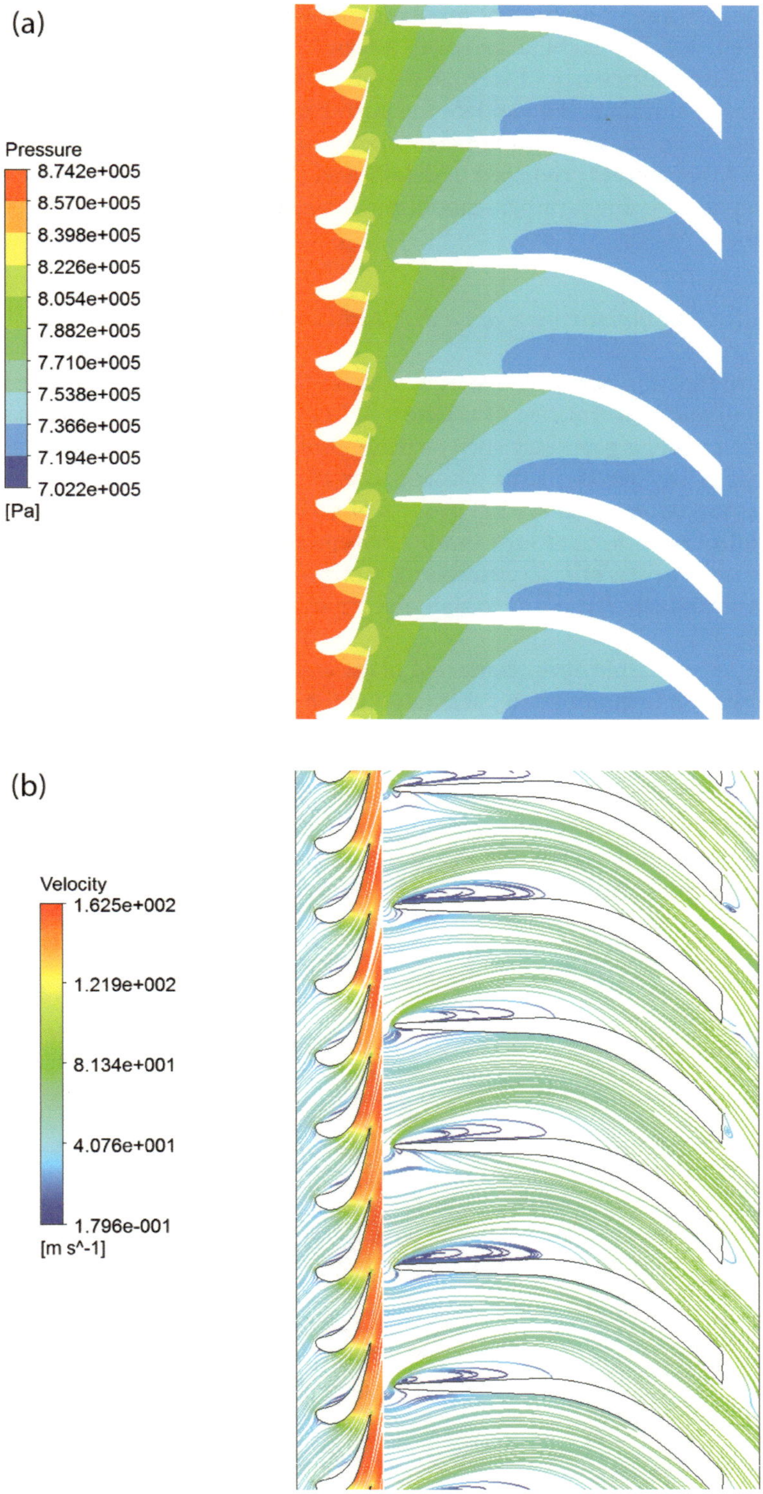

Figure 12.
Pressure contour and velocity streamline diagram of nozzle and impeller passage section at 50% spanwise. (a) Pressure contour, and (b) Velocity streamline diagram.

meeting the design requirement. The pressure contour shows a small vortex area of low pressure at the suction side of blade trailing edge, which is also seen in the velocity streamline diagram.

Therefore, through numerical simulation, the actual performance of turbine can be estimated, and the part that needs to be optimized can be figured out to make the aerodynamic efficiency of turbine better. In addition, numerical simulation can also be used to simulate and validate the performance of OTEC turbine in other aspects. For example, in [17], the performance of OTEC turbine under design and off-design conditions is predicted in detail. And the authors in [18] discussed the effects of turbine blade number, mounting angle, and blade fillet on OTEC turbine performance.

2.3.3 Aerodynamic optimization research of turbine in OTEC

The aerodynamic performance of turbine in OTEC is significantly affected by the aerodynamic performance of nozzle and impeller. Therefore, aerodynamic optimization of flow geometry such as blade profile is crucial to improve flow characteristics and aerodynamic performance of turbine.

The aerodynamic optimization process of turbine is a reciprocating process of design modification, evaluation, and improvement. It mainly includes selecting design variables, determining and solving optimization objectives, and establishing the relationship between optimization objectives and design variables, as shown in **Figure 13**.

Before aerodynamic optimization of blade cascade, it is necessary to use some curve expression technology to parameterize the geometry of optimized object and determine the design variables, such as NURBS curve, Bézier curve, and nonrational B-spline methods. Then, the numerical software is used for geometric modeling, meshing, numerical solution of flow field, and flow field treatment and analysis after solving, so as to analyze and evaluate the aerodynamic performance of the optimized turbine and obtain the objective function value. Finally, the optimal design algorithm is used to search for the optimal design parameters, and then optimization parameters are used to improve the design variable parameters, so that the optimal solution can be quickly searched and the optimization process automation can be realized.

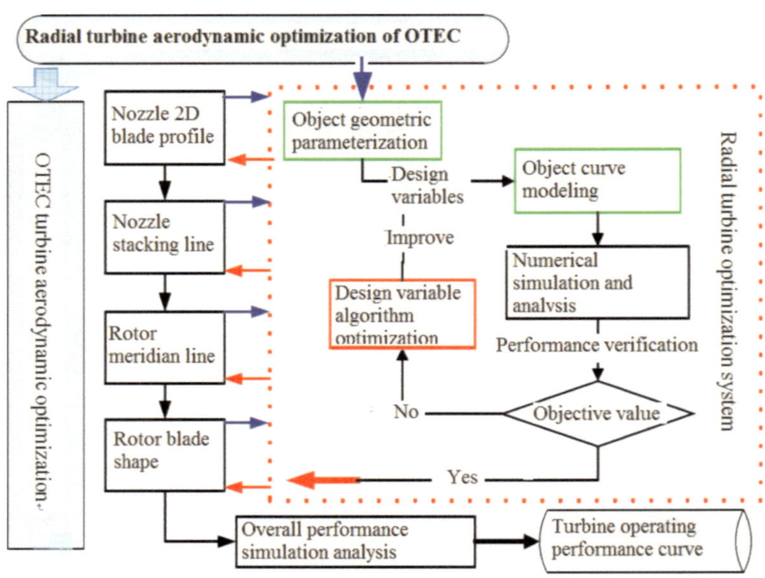

Figure 13.
Optimization process of radial turbine.

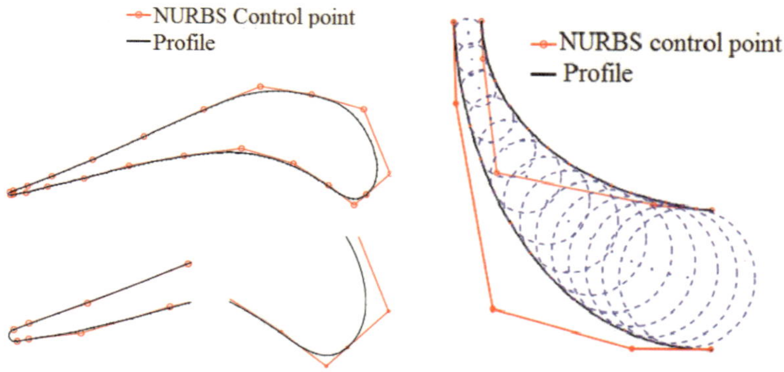

Figure 14.
Profile of nozzle blade and impeller meridian blade.

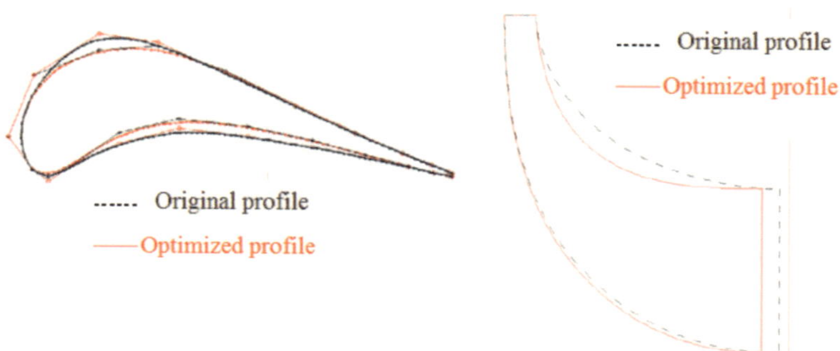

Figure 15.
Original and optimization profile of nozzle and impeller blade.

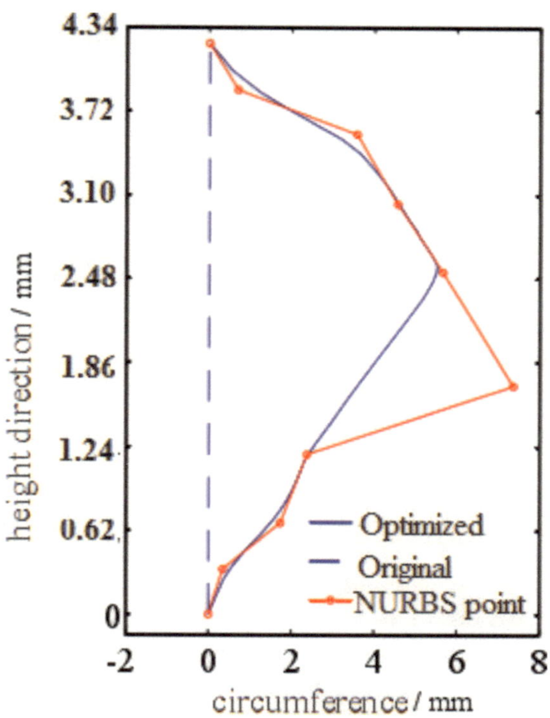

Figure 16.
Optimization of nozzle stacking line.

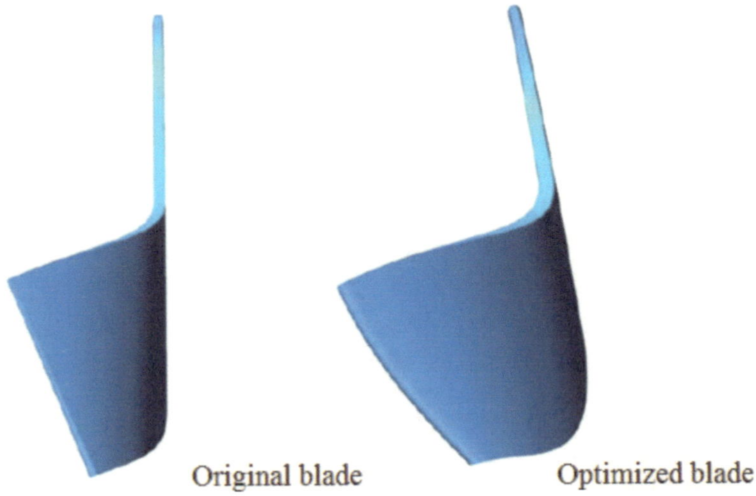

Figure 17.
Original and optimized blade of impeller.

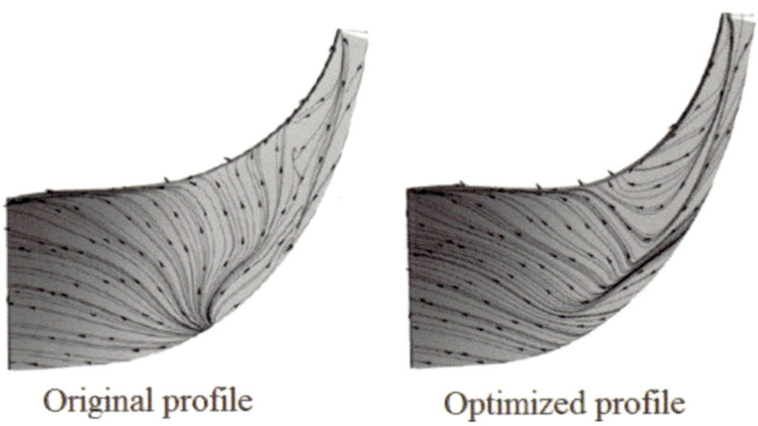

Figure 18.
Flow line at the suction surface of impeller.

In this case, NURBS curve is used to control nozzle blade profile and meridian profile of impeller, as shown in **Figure 14**. By integrating the above optimization method with ANSYS CFX program, geometric modeling, meshing, numerical solution, and performance analysis of flow field are carried out. And then the optimization algorithm is used to improve the designed variable parameters to search for the optimal solution quickly. Finally nozzle blade profile and impeller blade profile are obtained, as shown in **Figure 15**.

In addition to using control points to optimize nozzle and impeller, in order to ensure the design conditions and meet the performance requirements, the aerodynamic performance of cascade stacking line is optimized by "sweeping" method considering that nozzle is a radial cascade. The optimized stacking line is shown in **Figure 16**. For impeller blades, the backward-curved blades shown in **Figure 17** are used to improve the flow characteristics of turbine impellers further.

Finally, a comparative analysis of optimized nozzle and impeller performance is performed. **Figure 18** shows the streamline at the suction surface of impeller. It can be seen that optimized meridian channel curve changes uniformly, and the flow direction is controlled to change more gently, thus eliminating partial separation of

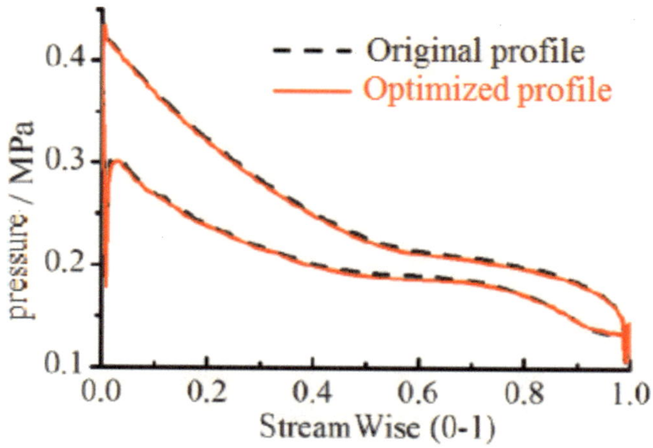

Figure 19.
Static pressure curve at 50% spanwise of impeller surface.

suction surface in impeller. The static pressure curve at 50% spanwise of impeller surface shown in **Figure 19** indicates that the static pressure is in the changes of favorable pressure gradients. The improvement of the static pressure change of blade suction surface can eliminate the local flow separation at suction surface of the original impeller. **Figure 20** shows the optimized streamline of nozzle and impeller at 50% spanwise. Comparing with **Figure 12**, it can be seen that the optimized flow field is more uniform without local vortex. Finally, the operation performance of turbine is numerically calculated using the optimized nozzle profile and impeller profile. The results show that operation efficiency of turbine is 1.05% higher than the original under design conditions.

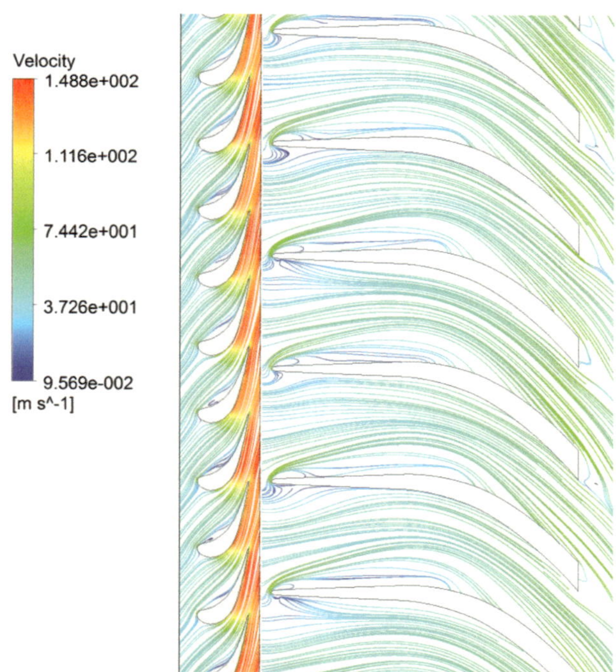

Figure 20.
Optimized streamline diagram of nozzle and impeller at 50% spanwise.

3. Mechanical and electric control design of turbine

This part contains the brief design of mechanical system and electric control system of turbine. In terms of mechanical design, in addition to the vibration and strength of impeller, the sealing design to prevent leakage of working fluid is also an important part of the research. Finally, in order to ensure the stability and safe operation, control part of turbine will also be introduced as the design content.

3.1 Mechanical design of turbine

3.1.1 Mechanical design of impeller

After the aerodynamic design of turbine is finished, mechanical research on material, strength, and vibration needs to be carried out in order to achieve the efficient and stable operation of turbine. Since working fluid is corrosive and turbine has high rotating speed, it is necessary to control the manufacturing process of blades. Therefore, requirements for impeller blade material of turbine include mechanical fatigue resistance, sufficient plasticity and toughness, better chemical stability with higher corrosion resistance, better vibration damping, superstructural manufacturability, and so on. Aluminum alloys, stainless steels, and titanium alloys are usually used as impeller materials according to the requirements of environment and economy.

In addition, the assembly and balance of rotating components are closely related to the safe operation of turbine. Therefore, in order to make turbine safe and stable, impeller must be assembled to have a specified axial clearance and radial clearance with respect to components such as nozzle ring and turbine cylinder. Before impeller is assembled, strict static balance must be carried out to check whether mass distribution is uniform, and unbalanced mass should be eliminated or reduced to permissible range. After assembly, dynamic balance of the whole rotating parts including impeller and shaft should be carried out to ensure the balance accuracy of dynamic parts.

Finally, the strength check of impeller is an indispensable part of turbine design. If the designed impeller does not meet the strength requirements, the aerodynamic scheme must be redistributed until the strength is satisfied. In this part, ANSYS is used to analyze the strength of impeller and rotor in terms of stress, strain, and vibration mode and check whether safety factors meet the requirements. The simulation of impeller strain is shown in **Figure 21**.

3.1.2 Mechanical structure design of turbine

In addition to pneumatic system, the turbine includes oil system, shafting system, sealing system, and power generation control system.

Oil system generally includes oil circuit for dynamic seal lubrication, oil circuit for bearing oil film, and oil circuit for generator heat dissipation. Generally, the first two parts have certain pressure, which not only provide lubrication and heat dissipation for sealing and bearing but also ensure that pressure is higher than pressure on working fluid side to have a certain sealing effect. The latter oil circuit is an ordinary oil system. The shafting system mainly includes the turbine main shaft, bearing, coupling, reducer, etc., which plays the role of transferring the work of impeller rotation to motor for power generation. The sealing system includes dynamic and static sealing parts. The dynamic sealing component is mainly used to

Figure 21.
Strain contour of turbine impeller.

seal flow channel system from external system at the shaft during the high-speed rotation of shafting. They have various structural forms, which will be introduced in detail in the next part. The static sealing component is mainly an O-ring that contacts the stationary components to isolate the working fluid, lubricating oil, and outside air. If high-speed generators are used for power generation, it can be driven by turbo-side shafting to generate electricity at a high speed and then converted into 50/60 Hz conventional power by rectification inversion. While the control part mainly controls turbine steam volume, speed, and generator operation, it also includes digital control system with emergency interruption, which will be described in the next section.

Figure 22 shows the schematic diagram of turbine. Mechanical seal is used for turbine in this case, and a high-speed generator is used for power generation. Therefore, the shafting system mainly includes the main shaft, high-speed bearing, and high-speed coupling. The main shaft is an important component connecting impeller and high-speed motor, which transfers the work of impeller to a high-speed coupling. There are two high-speed bearings, which play a supporting role in the main shaft rotating process and ensure that the shaft does not move in the circumferential and axial direction when rotating at a high speed. The high-speed coupling connects the main shaft and high-speed generator to ensure that the work of the main shaft can be smoothly transmitted to a high-speed generator. Oil circuit

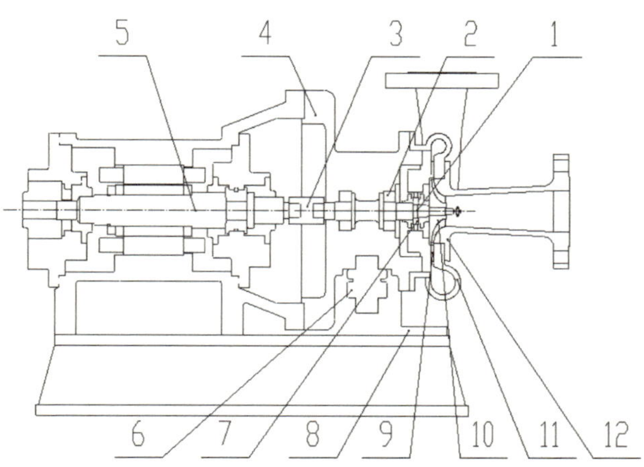

Figure 22.
Schematic diagram of turbine (1) spindle, (2) bearing, (3) coupling, (4) body, (5) high-speed generator, (6) oil pipe, (7) sealing, (8) plate, (9) nozzle, (10) impeller, (11) volute, (12) diffuser.

system includes three parts, oil circuit of sealing device, oil circuit of high-speed bearing, and oil circuit of high-speed generator, represented by oil supply pipeline. Lubricating oil passes through the above three parts and enters the tank for cooling and recycling.

The flow passage system is the passage for ammonia designed in the previous chapter, including volute, nozzle, impeller, and diffuser. The saturated ammonia vapor heated by warm seawater first flows into volute, so that the airflow is evenly distributed in circumferential direction. The airflow expands and accelerates after entering nozzle with a large kinetic energy and then enters impeller at a high speed to drive the main shaft. Finally, ammonia is discharged through diffuser. In addition, the body and plate mainly have the function of wrapping and supporting the turbine. Mechanical seal belongs to dynamic seal, and static seal is distributed in the connection of corresponding static components.

3.1.3 The sealing of turbine

Since ammonia or organic working fluid is used as working fluid in OTEC, leakage should be avoided as much as possible. Turbine is the device that drives high-speed motor or speed reducer to rotate at a high speed. Therefore, besides ensuring static sealing inside turbine, it is also necessary to ensure that shaft-end dynamic seal can prevent the leakage of working fluid as well. Sealing has become a key issue in turbine design, as well as the design of shaft sealing. The shaft-sealing device of turbine is a kind of dynamic sealing device installed on the main shaft to minimize the leakage of working substance inside the turbine to outside. There are generally two shaft-sealing devices inside the turbine. One is the shaft seal installed at the high-pressure end of turbine, which is mainly to prevent working fluid from entering the main shaft to corrode lubricating oil or bearing to cause bearing wear. The other is installed at the connection between the main shaft and motor, which is mainly to reduce the leakage of the working fluid to outside world and then to reduce the pollution to environment and harm to human.

The turbine shaft sealing mainly includes the following types: labyrinth seal, floating ring seal, carbon ring seal, mechanical seal, and dry gas seal. The principle of labyrinth seal is that when gas passes through the cogging gap, it obtains a certain speed. In the larger space between teeth, its kinetic energy is transformed into heat energy, while pressure decreases, and process of pressure reduction is repeated between a cogging gap. Although the labyrinth seal technology is very mature, it has the disadvantage of large leakage compared with other seals. Therefore, the labyrinth seal can be used as auxiliary seal, instead of directly sealing working fluid only with it.

Floating ring seal is mainly composed of an inner floating ring and an outer floating ring. The pressure between inner and outer floating rings is slightly higher than that of sealed working fluid. The inner and outer floating rings restrict internal and external leakage through small clearance throttling. The floating ring seal has the following advantages, long service life, high reliability, suitable for high-speed, wide operating range, etc., but it also has the disadvantages of large leakage and complex control system.

Carbon ring seal has structure which combines labyrinth seal and floating ring seal. According to the internal structure of carbon ring, it can be divided into flow resistance contact and non-contact sealing structures. The former has no gap between carbon ring and shaft and compensates itself with the wear of the carbon ring, while the latter has a working gap between carbon ring and the shaft. The selection of carbon ring seal structure is mainly based on the working conditions. It is widely used in steam turbines and compressors because of its convenient installation, good sealing, easy maintenance and repair, simple system, and no need of lubrication and cooling.

Mechanical seal is a leakage-preventing device which is formed by a pair of end faces perpendicular to the rotation axis, which keep sticking and merging relative sliding under the action of elastic force of compensation mechanism and the cooperation of auxiliary seal. The main component of a mechanical seal is a seal ring consisting of a moving ring and a static ring. The mechanical seal mainly has the advantages of reliable sealing, stable sealing state in long-term operation, low power consumption, good vibration resistance, and wide application range. However, since the mechanical seal is internally matched to improve the sealing performance, the main disadvantages are that the structure is complicated, it includes high processing requirements, installation and replacement are troublesome, and the cost is high.

The structure of dry gas seal is similar to that of mechanical seal. It is also composed of static ring, moving ring, spring, auxiliary seal ring, and transmission mechanism of moving ring, and there is a hydrodynamic groove on the outer side of moving ring. When the shaft of the main engine rotates, the ring rotates, and then the gas in the sealing chamber enters the sealing end face along the hydrodynamic groove. Dry gas seals can be divided into single-end seals, tandem seals, tandem seals with intermediate labyrinth seals, and double-face seals. The dry gas seal has the characteristics of small leakage, low power consumption, and long service life and is suitable for sealing of toxic and harmful working fluids.

Considering the main features of various seals and high sealing requirement due to high speed of turbine, dry gas seal and mechanical seal are suitable for the turbine if improved well. For example, the turbine in this case adopts the improved double-end mechanical seal structure, which ensures zero leakage and stable operation of turbine. In addition to dynamic seal at the shaft end, for static seal, such as the seal at inlet and outlet, the corrosion-resistant O-ring is chosen to seal on the joint surface.

3.2 Control system design of turbine

The control system of turbine in OTEC power generation system mainly includes a speed regulating system and automatic shutoff system, as shown in **Figure 23**. Speed regulating system changes the inlet flow rate by controlling the opening of the control valve, so as to make turbine work as required. Auto shutoff system will shut down the intake valve quickly in emergency to protect the safety of unit. Digital electro-hydraulic (DEH) system is referred to as speed regulation system, and DEH system realizes the functions of program-controlled start-up, automatic adjustment, parameter limitation, protection, and monitoring of turbine by control loops working separately or jointly.

The automatic adjustment control function is mainly to set the turbine target speed; the unit can automatically control the regulating valve along empirical curve corresponding to current state and complete speed-up control until target speed. In the process of speedup, the operator can also control the speed-up process by modifying the target speed, rate of increase, and speed-keeping time.

When the unit dumps load, DEH accepts oil switch tripping and overspeed signal and quickly closes the regulating valve to reduce the conversion overshoot. The valve will open automatically after a delay or recovered normal speed and maintain unit speed as target speed. This process can be achieved by limiting control function.

Emergency manual means that when DEH communication fails, the automatic control will automatically or manually turn to emergency manual state. The control valve can be adjusted and reduced according to the hard disk, and the valve opening can be controlled directly through the servo plate.

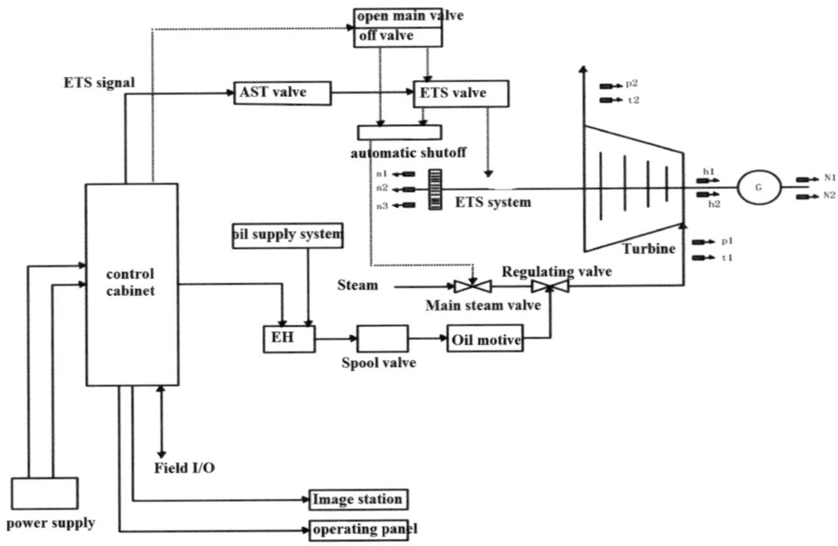

Figure 23.
Schematic diagram of turbine control system.

Overspeed protection means that when the speed exceeds the tripping speed, DEH sends out a signal to block the system action and quickly closes the main steam valve and regulates the steam valve.

In order to ensure the reliability of DEH system, system design should conform to safety design principles stipulated by the International Electrotechnical Commission (IEC). There should be preventive measures for possible operations, and turbine can be stopped safely when power source is lost. The whole control system should be dominated by electric control and the hard hand control is used as a backup. Redundancy design should be adopted in the whole system, and strict tracking measures should be taken to make switching between different modes without disturbance.

Emergency Trip System (ETS) for high-speed turbine needs to realize the following functions: continuously monitor turbine operation safety, and jump out in time to ensure safety in case of emergencies; the anti-interference performance complies with relevant standards to avoid interference of external radio signals leading to wrong operation; alarm signal uses sound-light alarm; and independent control processing module and measurement components are used to ensure reliability. When the following happens, turbine overspeed to tripping speed, thrust pad temperature is high, lubricant pressure is low, axial displacement is large, bearing vibration is large or manually shut down, etc., ETS will issue a jump instruction. In order to ensure the normal operation of bearings, it is necessary to monitor the temperature of bearings and lubricants as well as oil pressure of lubricants. In control system, turbines must be equipped with auxiliary system monitoring and protection functions as well as low oil pressure protection devices.

In addition, in order to ensure that the turbine control system has good dynamic response characteristics and sufficient valve lift force, it is necessary to have electro-hydraulic (EH) oil system to provide control and power oil for the turbine servo system and protection system. Lubricating system needs to provide lubricating oil and sealing oil for turbine bearings and high-speed generators and pressure oil for mechanical overspeed emergency interruption system.

4. Problems and prospect of turbine in OTEC

This section mainly introduces the problems in the turbine currently and the prospects for the development of the turbine in OTEC.

4.1 Technical issues of turbine in OTEC

There are technical issues to be resolved in the turbine systems currently running for low efficiency, leakage, operation stability, etc. So far, the authors have found the following technical difficulties of turbine in OTEC through research for several years:

1. At present, the design of turbine is not initiated for OTEC processes, but basically follows conventional radial turbine design methods. The selection of empirical parameters, calculation of geometric parameter, and 3-D modeling are all referenced to the turbines used in waste heat power generation and low-temperature power generation. There are barely any specific aerodynamic designs and performance studies for the small temperature difference of turbine, specifically for OTEC. Therefore, the aerodynamic efficiency varies with research methods applied, and therefore it may not be promising to reach the highest efficiency in OTEC.

2. Ammonia or organic working fluids used in the system have potential impacts on the environment and humans, especially because ammonia is toxic and irritating and may explode under unexpected thermodynamic conditions. Unlike other parts of the OTEC system such as heat exchangers and pumps, the turbine rotates at a high speed, so that the turbine sealing should be perfectly guaranteed for long-term operations of OTEC. The seals introduced above often have low performances and sophisticated internal structure so that it is hard to replenish while the OTEC system is on and only remote control is available. Therefore, the zero leakage of working fluid in turbine is an everlasting key issue to be addressed.

3. The power generated by OTEC system is relatively low, as low as 100 kW. In other words, the high-power systems are built at small-scale systems during R&D phases, having the high cost and slow recovery of benefits. As high-power generation systems are rare, the high-power turbine for OTEC is also one of the future development tasks.

4.2 Prospect of turbine in OTEC

Looking for solutions in view of the current problems of power generation turbines, this section looks to the future to find development direction of the turbine in OTEC. For the problems existing in turbines for OTEC, the authors have opinions that the turbines can be improved by addressing the following aspects of turbine design and manufacturing in the future:

1. The magnetic bearing and drive technology currently used in other rotating machinery can be adapted in the turbine of OTEC. Although magnetic bearing has the characteristics of little friction loss and high-speed performance, its direct application to OTEC turbine requires maturity of its own technology at a low cost. Magnetic drive technology is currently mainly used in high-power,

low-speed rotating machinery, which can completely replace the ordinary coupling to ensure zero leakage of working fluid. But due to its low transmission efficiency, it is necessary to ensure that it's suitable in high-speed turbines without hysteresis and great loss in the future.

2. In order to improve the aerodynamic performance of turbines in OTEC, it is necessary to seek a special optimization strategy for the blade profile of the turbines to optimize the aerodynamic part of the turbine. So as to ensure the maximum efficiency of the OTEC system, it's necessary to improve the aerodynamic efficiency of the turbine to the highest efficiency in OTEC.

3. In order to achieve the operation of the high-power OTEC system as soon as possible, it is necessary to have research on the performance of high-power turbine in which the main directions are low speed, turbine type, and so on.

5. Conclusions

This chapter mainly introduces the turbine in OTEC, including the design method of turbine such as aerodynamic, mechanical, and control design. Finally, the current problems and prospects of the power generation turbine in OTEC are discussed. It should be noted that the design method of turbine is not limited to the method introduced in the chapter. The authors hope that the readers can have a comprehensive and brief understanding of the turbine in OTEC through the introduction of this chapter. On this basis, the readers can also be expected to conduct a more in-depth study and discussion of the turbine in OTEC.

Acknowledgements

This work was supported by the China-Korea Cooperation Project "China-Korea Technology Exchange and Cooperation of Ocean Energy Development and Utilization" (PI-2018-4).

Nomenclature

Symbols

$\overline{D}_2$	wheel diameter ratio
$\overline{u}_1$	velocity ratio
c	absolute velocity
u	circumferential velocity

Greek symbols

η	efficiency
Ω	degree of reaction
ω	relative velocity
α	absolute flow angle
β	relative flow angle
φ	velocity coefficient of impeller
ψ	velocity coefficient of nozzle

Subscripts

j = 1, 2	inlet and outlet of impeller
u	circumferential direction
c	cold seawater
i	inlet
o	outlet

Author details

Weimin Liu[1*], Yunzheng Ge[1], Lei Liu[1] and Yun Chen[2]

1 First Institute of Oceanography, Ministry of Natural Resources (MNR), Qingdao, China

2 Shandong University, Qingdao, China

*Address all correspondence to: lwmxjtu@163.com

References

[1] Schobeiri T. Thermo-fluid dynamic design study of single and double-inflow radial and single-stage axial steam turbines for open-cycle thermal energy conversion net power-producing experiment facility in Hawaii. Journal of Energy Resources Technology. 1990; **112**(1):41-50. DOI: 10.1115/1.2905711

[2] Yang MH, Yeh RH. Analysis of optimization in an OTEC plant using organic Rankine cycle. Renewable Energy. 2014;**68**(3):25-34. DOI: 10.1016/j.renene.2014.01.029

[3] Yoon JI, Son CH, Baek SM, et al. Efficiency comparison of subcritical OTEC power cycle using various working fluids. Heat and Mass Transfer. 2014;**50**(7):985-996. DOI: 10.1007/s00231-014-1310-8

[4] Uehara H, Ikegami Y. Optimization of a closed-cycle OTEC system. Trans of the ASME Journal of Solar Energy Engineering. 1990;**112**(4):247-256. DOI: 10.1115/1.2929931

[5] Wu C. Specific power optimization of closed-cycle OTEC plants. Ocean Engineering. 1990;**17**(3):307-314. DOI: 10.1016/0029-8018(90)90007-s

[6] Han S, Seo JB. 20kW turbine development for OTEC system. The KSFM Journal of Fluid Machinery. 2014; **17**(6):38-43. DOI: 10.5293/kfma.2014.17.6.038

[7] Chen F, Liu L, Peng J, et al. Theoretical and experimental research on the thermal performance of ocean thermal energy conversion system using the Rankine cycle mode. Energy. 2019; **183**:497-503. DOI: 10.1016/j.energy.2019.04.008

[8] Rama SRG, Aijaz AK. Turbomachinery: Design and Theory. Boca Raton: CRC Press; 2003

[9] Whitfield A, Baines NC. Design of Radial Turbomachines. Harlow: Longman Scientific Technical; 1990

[10] Balje OE. Turbomachines: A Guide to Design, Selection and Theory. New Jersey: Wiley; 1981. DOI: 10.2307/27851419

[11] Whitfield A. The preliminary design of radial inflow turbines. Journal of Turbomachinery. 1990;**112**(1):50. DOI: 10.1115/1.2927420

[12] Costall AW, Gonzalez HA, Newton PJ, et al. Design methodology for radial turbo expanders in mobile organic Rankine cycle applications. Applied Energy. 2015:S0306261915002573. DOI: 10.1016/j.apenergy.2015.02.072

[13] Rahbar K, Mahmoud S, Al-Dadah RK, et al. Modelling and optimization of organic Rankine cycle based on a small-scale radial inflow turbine. Energy Conversion and Management. 2015;**91**: 186-198. DOI: 10.1016/j.enconman.2014.12.003

[14] Rahbar K, Mahmoud S, Al-Dadah RK, et al. Parametric analysis and optimization of a small-scale radial turbine for organic Rankine cycle. Energy. 2015;**83**:696-711. DOI: 10.1016/j.energy.2015.02.079

[15] Rahbar K, Mahmoud S, Al-Dadah RK. Mean-line modeling and CFD analysis of a miniature radial turbine for distributed power generation systems. International Journal of Low-Carbon Technologies. 2014:ctu028. DOI: 10.1093/ijlct/ctu028

[16] Li Y, Ren XD. Investigation of the organic Rankine cycle (ORC) system and the radial-inflow turbine design. Applied Thermal Engineering. 2016;**96**: 547-554. DOI: 10.1016/j.applthermaleng.2015.12.009

[17] Nithesh KG, Chatterjee D. Numerical prediction of the performance of radial inflow turbine designed for ocean thermal energy conversion system. Applied Energy. 2016;**167**:1-16. DOI: 10.1016/j.apenergy.2016.01.033

[18] Nithesh KG, Chatterjee D, Oh C, et al. Design and performance analysis of radial-inflow turboexpander for OTEC application. Renewable Energy. 2016;**85**:834-843. DOI: 10.1016/j.renene.2015.07.018

Chapter 4

Novel OTEC Cycle Using Efficiency Enhancer

Hosaeng Lee, Seungtaek Lim, Jungin Yoon and Hyeonju Kim

Abstract

The ocean thermal energy conversion (OTEC) plant is designed to improve the efficiency of the existing plants. Various researches are being conducted to increase the plant's efficiency and output with the use of an enhancer, and studies for performance improvement are also in progress from the Kalina and Uehara cycles to ejector pump OTEC (EP-OTEC). Their performance can be improved by increasing the evaporation pressure using an unused heat source and reducing the heat consumption using a reheating system and a regenerator. In the case of EP-OTEC, an ejector is installed near the turbine-exit to reduce the pressure and therefore increase the power output. In simulations and experiments conducted in this study, EP-OTEC showed 38% efficiency improvement from the basic cycle, which is due to the power output volume increase. The optimum ratio was derived by adjusting the pressure ratio. The demonstration plant to be developed in the future is expected to be applied to the high-efficiency OTEC demonstration plant with improved performance, and new technologies will be continuously developed considering economics and commercial viability.

Keywords: enhancer, reheater, regenerator, ejector pump OTEC, motive pump

1. Introduction

Ocean thermal energy conversion (OTEC) processes run under a much lower temperature gradient than that of the conventional thermal engines. The actual efficiency of a heat engine driven by the typical temperature difference of 20C is much smaller than the theoretical maximum due to heat and friction losses in the OTEC system. As such, from this perspective, OTEC may appear as a viable alternative energy source only after its thermal efficiency is well improved. As surface seawater is used as a heat source, which is essentially available in large quantities as an alternative energy source, many evaluators consider OTEC as one of the likely renewable energy-source candidates for large-scale power generation, especially in the tropical regions.

To improve the low-efficiency OTEC process, research has been focused on increasing their operational efficiency by using a high-quality heat source and using extra devices capable of improving the system's performance, in addition to locating and operating OTEC plants in tropical areas. In 1981, the Kalina cycle, named after its inventor, was reported as a new method for efficient thermal energy conversion to overcome some disadvantages of the well-known closed Rankine cycle. Whereas the Rankine cycle requires pure substances like ammonia, the Kalina cycle can use a mixture of ammonia and water [1, 2].

In 1994, the Uehara cycle was invented at Saga University to design and build a 4.5-kW power plant. The Uehara cycle includes a combined process of absorption and extraction, showing 1–2% better performance than that of the Kalina cycle. As shown in **Figure 1**, Bluerise in the Netherlands is conducting basic experiments on OTEC employing the Kalina cycle. As shown in **Figure 2**, a 4.5 kW-class experimental plant using the Uehara cycle was tested [3–6].

Recently, Amyra et al. developed a solar-collected OTEC (SC-OTEC) cycle of solar thermal energy. Their heat-collection system exchanges heat by utilizing surface water of a high temperature for evaporation and deep seawater of a low temperature for condensation to improve the power generation performance. In addition, Hakan Aydin suggested a performance change by increasing the superheating degree of the OTEC plant through the use of a solar collector [7, 8].

In 1982, Kyushu Electric (Japan) succeeded in constructing a 50-kW OTEC plant of a closed-loop cycle, utilizing waste heat from a diesel generator. In 2015, a 200-kW high-temperature power generation plant was built by KRISO (Korea),

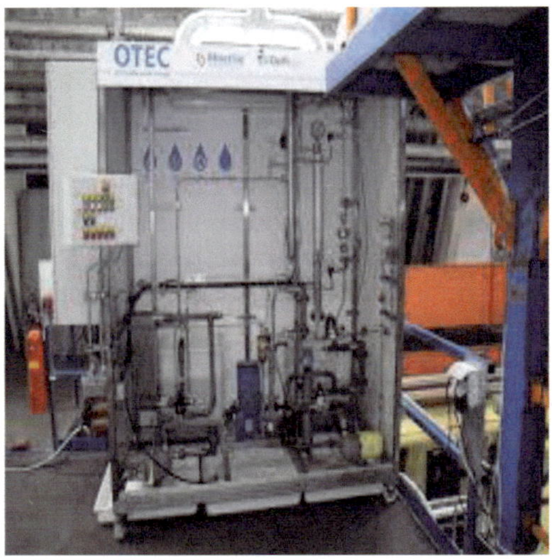

Figure 1.
Experimental equipment of OTEC using the Kalina cycle (Bluerise).

Figure 2.
4.5-kW experimental equipment of OTEC using theUehara cycle (Saga Univ).

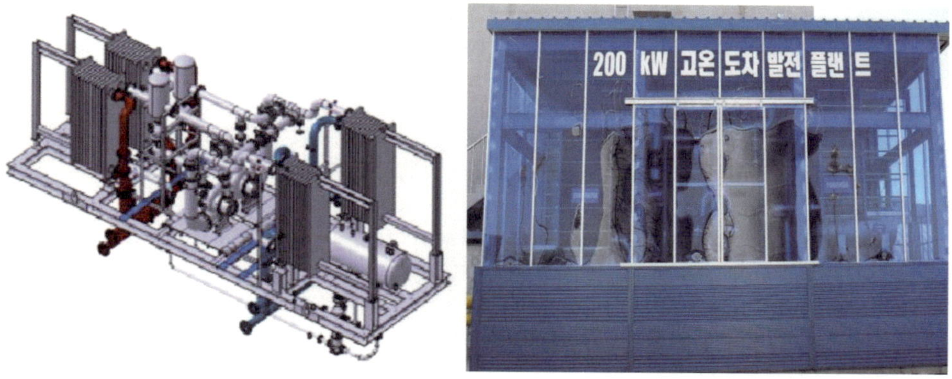

Figure 3.
A 200-kW HOTEC plant using 80°C heat source (KRISO): 3D graphic (left) and photographic image (right).

which taps 70–80°C industrial waste heat and cold deep sea water, as shown in **Figure 3**. The plant improved the power generation efficiency by 7.8% through manufacturing components and a series of experiments, laying the foundation for the industrialization of the OTEC plant that taps external heat sources [9].

In addition to efficiency improvement, research has been carried out for heat reduction through the use of a reheating system and a regenerator. Recently, the power output and efficiency were additionally improved by Ejector and Pump-OTEC devised by Yoon et al., producing the maximum generation efficiency of 3.1% at the surface seawater temperature of 26°C [10].

2. Theory for OTEC cycle

OTEC plants generate electric power using the temperature difference between deep sea water and surface water. The system efficiency of OTEC can be expressed as the energy production rate of a turbine per thermal energy from evaporator. As the temperature of the heat source from the surface seawater increases, the vapor pressure of the seawater increases, the enthalpy increases from the inlet to the outlet of the turbine, and hence the process efficiency and power output are

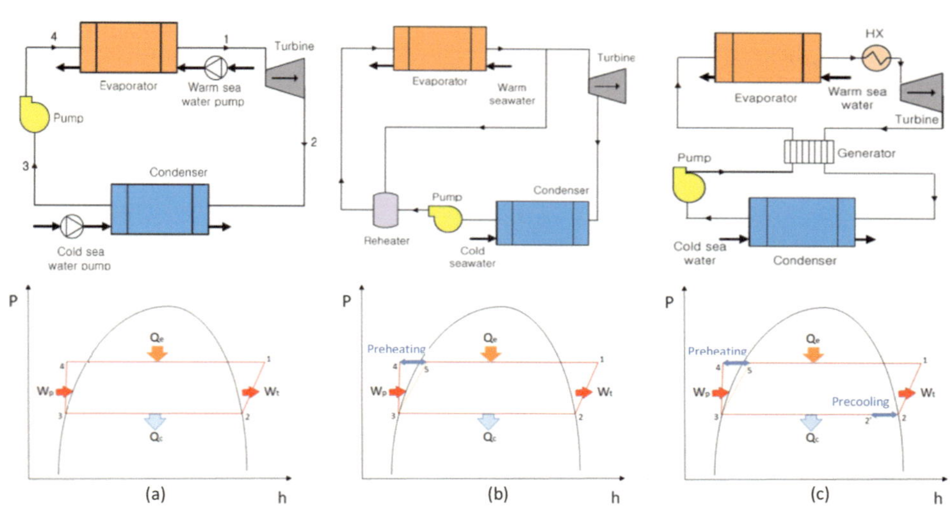

Figure 4.
Schematic of basic and novel OTEC cycle: (a) basic cycle, (b) using reheater, (c) using regenerator.

enhanced. The use of high-temperature heat sources using collectors and heat storage tanks that can recover solar, geothermal, and industrial waste heat can further increase the process efficiency. On the other hand, one can increase the efficiency of the OTEC system by reducing the required amount of heat consumption in the evaporator. Toward this end, various studies have suggested methods to improve the efficiency of the closed OTEC cycle. One such way is by attaching a regenerator and a reheater.

The novel cycle that improves the efficiency of OTEC is transformed from the basic Organic Rankine Cycle. The basic ORC cycle repeats the process of reversible adiabatic compression of working fluid pump, constant pressure heat addition in evaporator, reversible adiabatic expansion in turbine, and constant pressure heat rejection in condenser. The OTEC using the basic Rankine cycle is shown in **Figure 4(a)** using the following balance equations:

$$Q_e = \dot{m}_r(h_1 - h_4) \tag{1}$$

$$W_t = \dot{m}_r(h_1 - h_2) \tag{2}$$

$$Q_c = \dot{m}_r(h_2 - h_3) \tag{3}$$

$$W_p = \dot{m}_r(h_3 - h_4) \tag{4}$$

where Q_e and W_t are the evaporation capacity and power output of turbine of the OTEC plant, respectively, and $\mathrm{Q_c}$ and $\mathrm{W_p}$ are the condenser capacity and the output of the working fluid pump, respectively.

On the other hand, in the case of the OTEC cycle with a reheater designed to reduce the evaporation capacity. The reheater improves the cycle efficiency by transferring heat to the working fluid flowing into the evaporator by recovering the heat of the superheated steam at the turbine outlet. The cycle is shown in **Figure 4(b)**. In the case of the method incorporating a regenerator, a part of the working fluid vaporized in the evaporator is bypassed, and the heat exchanged with the working fluid flows into the working fluid pump, as shown in **Figure 4(c)**. Then, the temperature of the working fluid flowing into the pump increases to reduce the evaporation capacity, thereby improving the cycle efficiency.

The equation for calculating the evaporation capacity with the regenerator or the reheater is changed to

$$Q_e = \dot{m}_{ww}(h_1 - h_5) \tag{5}$$

While on the one hand the cycle efficiency increases due to the reduced evaporation capacity, on the other hand, for the reheater case, the turbine output is represented as

$$W_t = (\dot{m}_r - \dot{m}_{rh})(h_1^{\prime} - h_2) \tag{6}$$

because the mass flow rate of the working fluid flowing into the turbine is reduced by the amount supplied by the reheater. Here, $\dot{m}_{rh}$ represents the mass flow rate of the working fluid flowing into the turbine.

3. Novel OTEC cycle using ejector

3.1 Concept

To improve the performance of OTEC plants, research on efficiency improvement employing an ejector has been conducted. As mentioned earlier, the OTEC system using an ejector pump is called "EP-OTEC." As shown in the system schematic in **Figure 5** and the P-h diagram in **Figure 6**, some of the working fluid branched from the distributor installed at the inlet of the evaporator exchanges heat with the surface water of the ocean, which is a high heat source in the evaporator, and assumes a high-pressure gas state. This high-pressure working fluid drives the turbine and draws it into the suction section of the ejector through the pressure difference generated in the cycle. On the other hand, the remaining working fluid divided in the distributor is moved by the operating pump to the operating part of the ejector in the liquid state, without passing through the evaporator. At this time, the pressure rises according to the expected degree of lifting of the operating pump. This pressure rise is an important factor in the cycle because lower inlet pressure can be achieved when the actuator fluid is supplied with a stronger force to the nozzle of the liquid-vapor ejector. The lower inlet pressure in the OTEC system with ejectors leads to an increase in the pressure difference between the turbine inlet and outlet, which results in enhanced turbine performance. The operating fluid injected

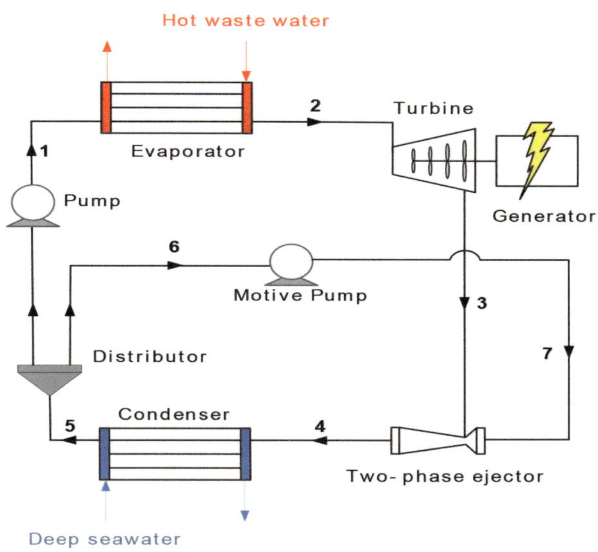

Figure 5.
Schematic of a two-phase ejector OTEC cycle.

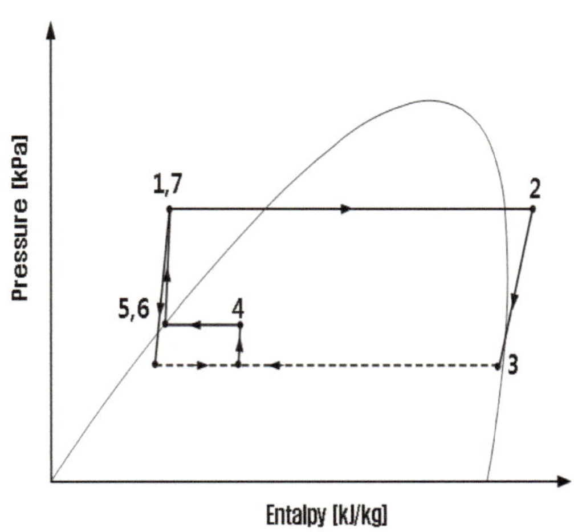

Figure 6.
P-h diagram of the EP-OTEC cycle.

from the ejector and the attracted suction fluid are mixed and discharged at the condensing pressure. The condenser condenses the working fluid after exchanging heat with deep water, which is a cooling source of the ocean. Finally, the working fluid is circulated by the pump, completing the full cycle.

The total power and net power of the turbine are determined by the mass flow through the turbine and the enthalpy change between the inlet and outlet of the turbine. Therefore, the enthalpy change due to the pressure change increased to generate more electricity using the turbine.

The power of the refrigerant pump is excluded from the turbine's total power equation (Eq. (7)), resulting in the turbine net power equation (Eq. (10)). A motive pump is newly applied to the current EP-OTEC system. The power input to the prime mover-pump is excluded from the turbine's total power because a high net power of the turbine is our goal as described using Eqs. (8) and (9). The system efficiency is expressed as the ratio between the input heat capacity and the output energy.

$$W_t = \dot{m}_r(h_2 - h_3) \tag{7}$$

$$W_{cp} = \dot{m}_r(h_1 - h_5) \tag{8}$$

$$W_{mp} = \dot{m}_r(h_7 - h_6) \tag{9}$$

$$W_{net} = W_t - W_{cp} - W_{\mathrm{mp}} \tag{10}$$

Figure 7(**a**) is a schematic diagram of an ejector pump and heat exchanger ocean thermal energy conversion (EPX-OTEC) cycle with an ejector pump and an actuator evaporator. The basic power generation principle is the same as that of the existing Organic Rankine cycle (ORC), but some working fluid is branched at the condenser outlet (after the boosting pump) before passing through the heat exchanger, exchanging heat with the used heat source and (partially) vaporizing it into the working part of the ejector. At this time, some vaporized working fluid serves to smooth the traction of the fluid at the ejector intake, and increases the volume of the working gas (gaseous phase of the working fluid) drawn into the ejector. The efficiency of the system varies with the dryness, optimization focuses on phase transformation. In the EPX and DEPX cycles, the actuating heat exchanger introduces a heat source with a reduced temperature after evaporating the working fluid. A low-pressure atmosphere is formed at the suction part by the traction force of the working fluid introduced into the ejector operating part, thereby reducing the pressure at the turbine outlet and obtaining more power generation at the same flow rate than in the basic cycle. The remaining fluid—branched out of the condenser—exits the evaporator, driving the turbine and entering the ejector.

Figure 7(**b**) is a schematic diagram of the dual ejector pump and heat exchanger (DEPX)-ORC with a dual ejector pump and an actuator evaporator. The basic principle of DEPX is the same as that of EPX, but in DEPX-ORC, the working fluid splits into two branches after the operation of the actuating pump, and some of it passes through the actuating evaporator, as in the EPX cycle, and then enters the actuating part of the primary ejector and pulling the fluid out of the turbine.

The EPX and DEPX cycles have the same system analysis and calculation principles as the EP-OTEC cycle, and both cycles contain an additional heat source called "actuating evaporator." Therefore, Eq. (11) expresses the amount of working fluid, branched by the P-h diagram and partial evaporation added by the amount of enthalpy before and after working-part evaporation, as shown in **Figure 7(b)**. The said equation is based on the EPX cycle, and the additional heat source is calculated in the same way as in the DEPX cycle.

$$Q_{e.extra} = \dot{m}_{r.mot}(h_7 - h_6) \tag{11}$$

Accordingly, the efficiency of the entire cycle is also changed such as,

$$n = \frac{W_{net}}{Q_e + Q_{e.extra}} \tag{12}$$

As shown in **Figure 7(c)**, the efficiency can be maximized through cascade OTEC, which uses a condenser and an evaporator on the high- and low-temperature sides, respectively. When a refrigerant with different pressures is used in the high/low-pressure part, low/high-pressure refrigerant is formed in the high/low-temperature part, and a low/high-pressure load is formed in the high/low-pressure part. Therefore, the high- and low-temperature sides were configured as DEPX-OTEC and EP-OTEC, respectively. Because it is difficult to apply the cascade

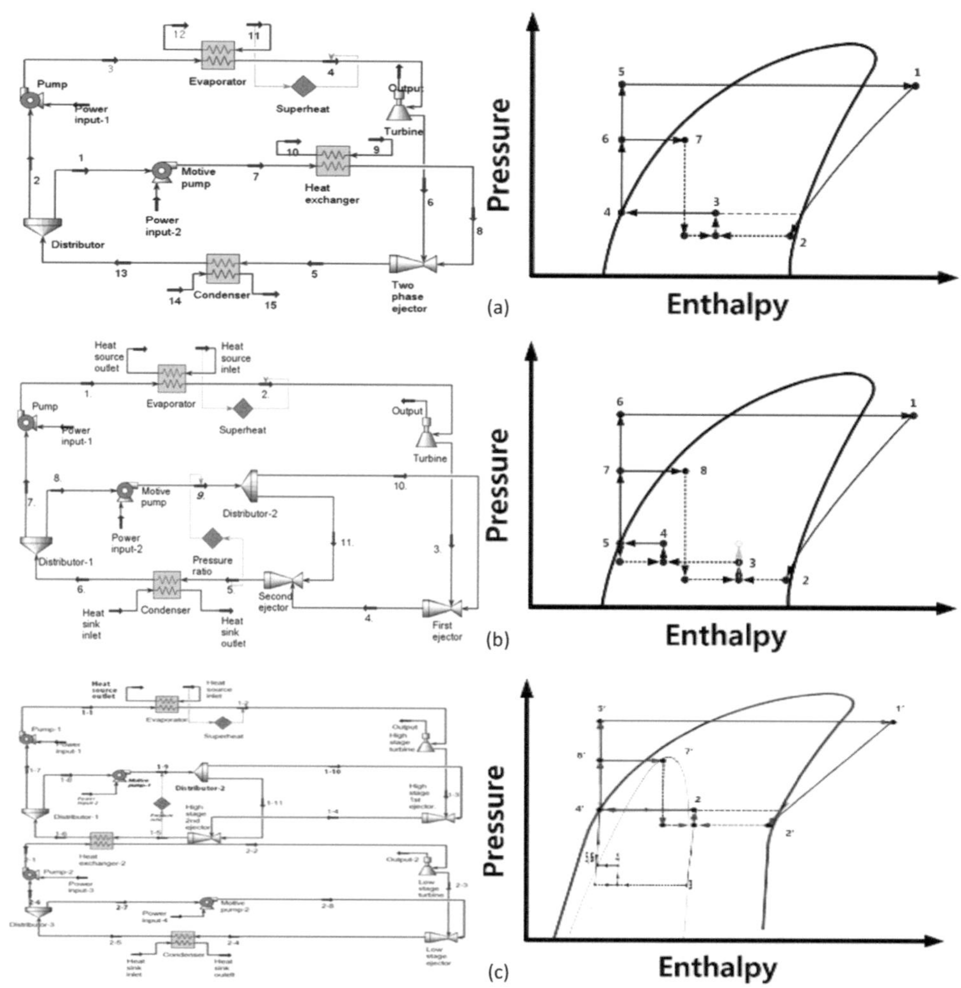

Figure 7.
Schematic of novel OTEC cycle using ejector: (a) EPX, (b) DEPX-ORC, (c) cascade DEP.

method to the existing OTEC cycles using low-temperature heat sources, the cascade method is suitable to employ unused heat sources for medium and low OTEC systems.

3.2 Simulation results

Table 1 shows the simulation conditions for the EP-OTEC system. The temperature of the marine surface water, which is a high heat source, was set to a typical temperature of the equator region of 29.0°C, which is almost invariant throughout the year. Deep seawater was applied to the condenser as the coolant, with the temperature kept about 5.0°C throughout the year.

The performances of conventional ocean thermal energy conversion and EP-OTEC were compared as the seawater temperature increased in the system with the

Parameters	Value
Surface seawater inlet temperature (°C)	24.0–29.0
Deep seawater inlet temperature (°C)	5.0
Temperature increase along the deep seawater intake pipe (°C)	5.8
Temperature increase along the surface seawater intake pipe (°C)	3.0
Pump efficiency (%)	65.0
Turbine efficiency (%)	80.0
Pressure ratio of motive and discharge of ejector (–)	3.0
Pressure drop at heat exchangers [kPa]	10.0

Table 1.
Analysis conditions.

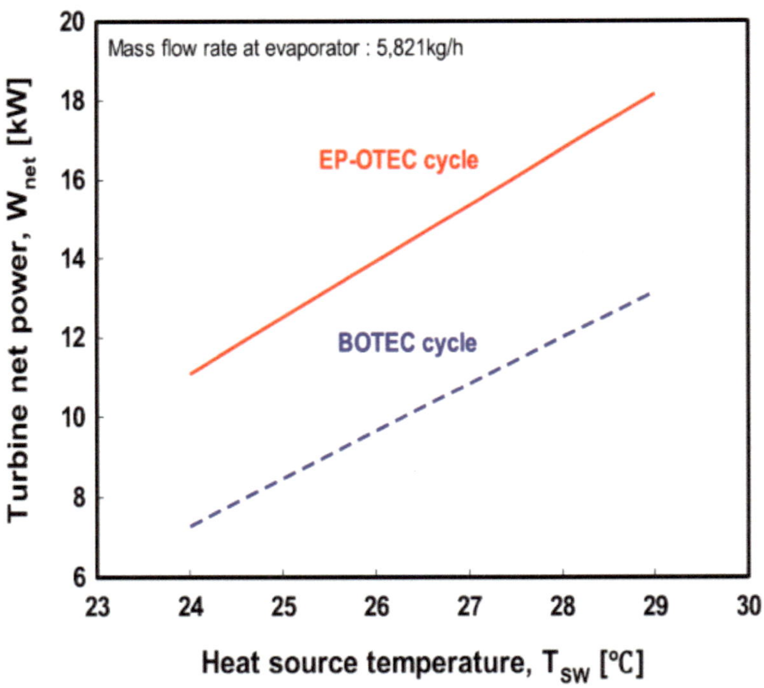

Figure 8.
Turbine net power with respect to the heat source temperature.

R152a refrigerant when the ejector operating fluid mass was kept constant [11]. **Figure 8** shows the net turbine output of the EP-OTEC cycle as the surface seawater temperature varied from 24.0 to 29.0°C. As the higher temperature, heat source entered the evaporator, evaporator pressure increases, resulting in an increase in

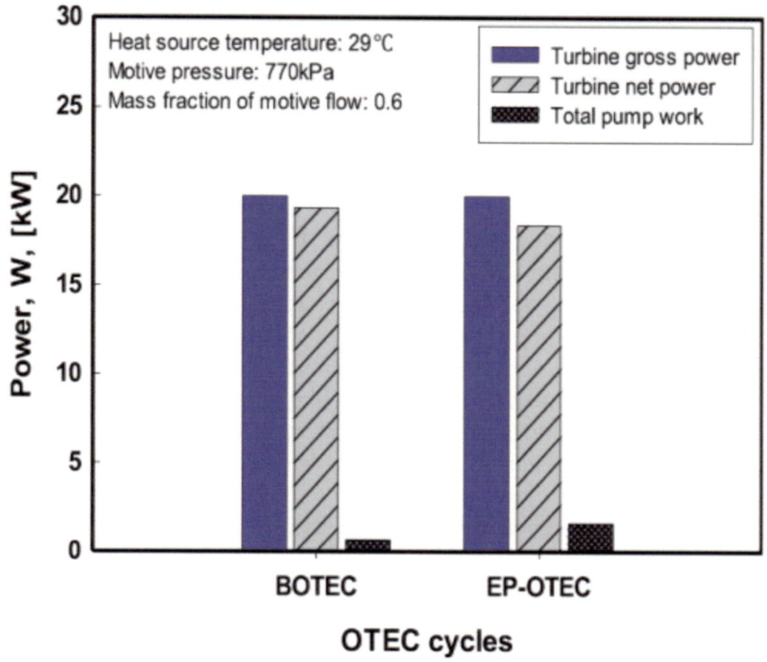

Figure 9.
Comparison of the turbine net power and the total pump work of the BOTEC cycle and the EP-OTEC cycle.

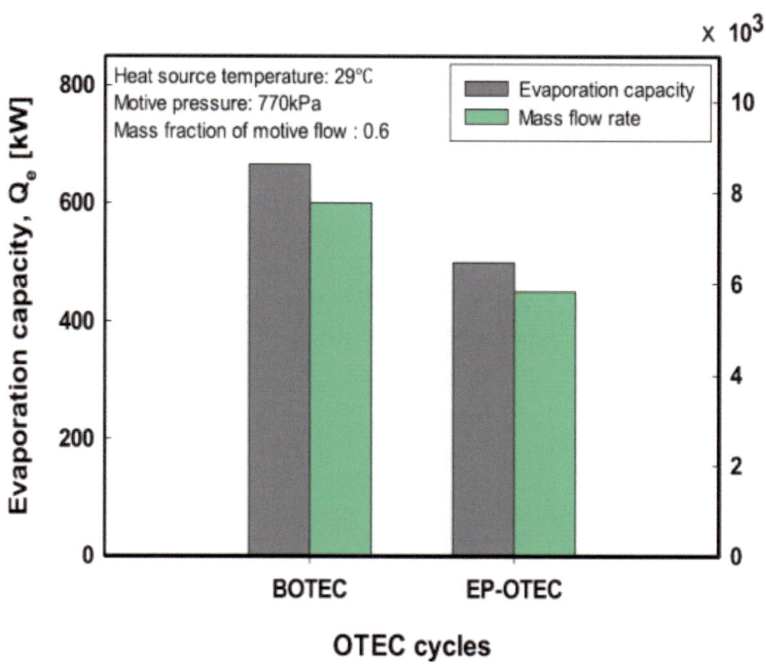

Figure 10.
Comparison of the evaporation capacity and the mass flow rate of the BOTEC cycle and the EP-OTEC cycle at the evaporator.

enthalpy from the inlet to the outlet of the turbine. For this reason, the turbine net power tended to increase with the surface seawater temperature at a constant mass flow rate to the evaporator. The utilization of the ejector and the motive pump lowered the turbine outlet pressure, and increased the enthalpy change of BOTEC and resulted in increased turbine power.

Figures 9 and **10** show that the net turbine output of the EP-OTEC cycle is lower than that of the BOTEC cycle, in spite of the advantages described above. This was due to the amount of mass flow used as the synchronous fluid of the ejector in the EP-OTEC cycle. About 40% of mass entering the evaporator becomes the suction fluid later in the process and the remainder is used as the motive fluid without even entering the evaporator. As a result, the evaporation capacity of the EP-OTEC cycle is about 33% lower than that of the BOTEC cycle at the same total turbine output.

Figure 11 compares the system efficiency of the EP-OTEC cycle with that of the BOTEC cycle for the 29°C surface seawater temperature and optimum conditions. The EP-OTEC cycle showed about 38% higher system efficiency than that of the BOTEC cycle, which confirms the superiority of the EP-OTEC to BOTEC.

In addition, comparative simulations of various cycles using ejectors were performed, employing unutilized heat sources like industrial waste heat. We designed an ORC engine using about 70–75°C temperature difference. In the heat exchanger of the ORC, the heat is exchanged with deep seawater heat sink and the low-temperature heat source below 80°C. Through the simulation, we compared the power generation and efficiency levels of EP, DEPX, and the cascade methods.

Figure 12 shows a graph comparing the efficiency of the BOTEC used in this study and those of the EPX and DEPX cycles. The EPX cycle and DEPX-ORC showed the reflected efficiency level when a flow rate ratio of 1.0 outside the pressure recovery rate of 1.4 was applied. Under the same heat source, heat sink, and power generation conditions, EPX-ORC showed 10.06 and 5% higher efficiency than those of those of the BOTEC when basic ORC exhibited 9.58% efficiency. In the case, using DEPX that is 11.07% efficiency, about 15% higher than that of the basic cycle, was achieved.

In the cascade cycle, R245fa and R152a with low saturation pressure were applied at the higher and lower stages, respectively. The difference between the

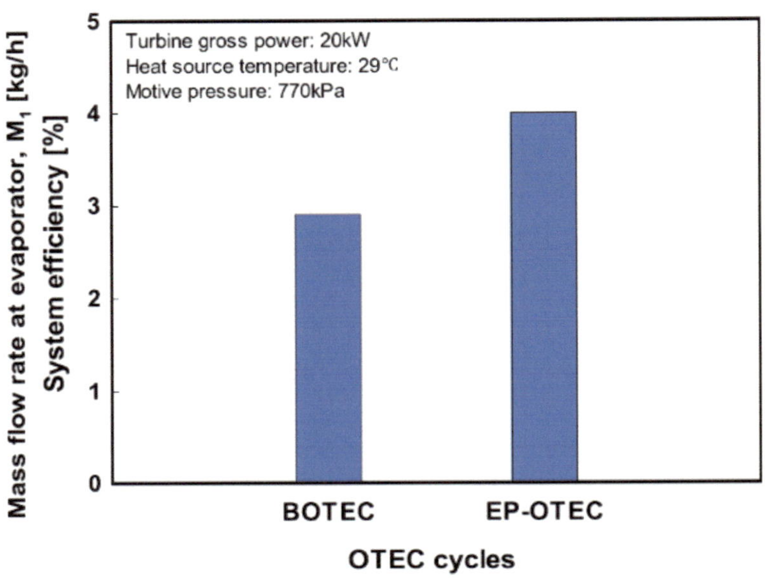

Figure 11.
Comparison of the system efficiency between BOTEC and EP-OTEC cycle.

evaporation and condensation pressures of R152a at 75°C warm water and 5°C cold water requires a high pump head about 1500 kPa, suggesting that R152a has a larger turbine inlet/outlet pressure difference than R245fa and R600a under the same temperature conditions. The cascade cycle was developed to enable secondary power production at a small temperature difference through its application at the

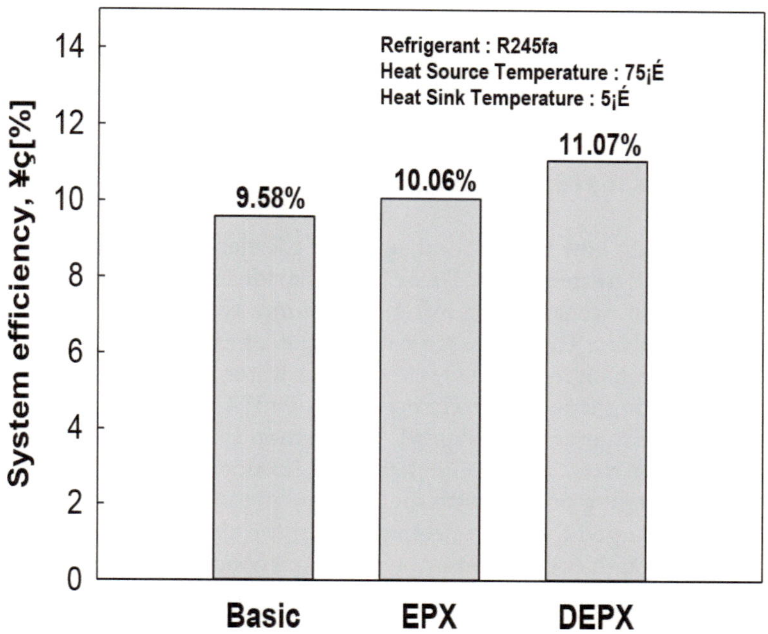

Figure 12.
System efficiency comparison with basic cycle between EPX and DEPX cycles.

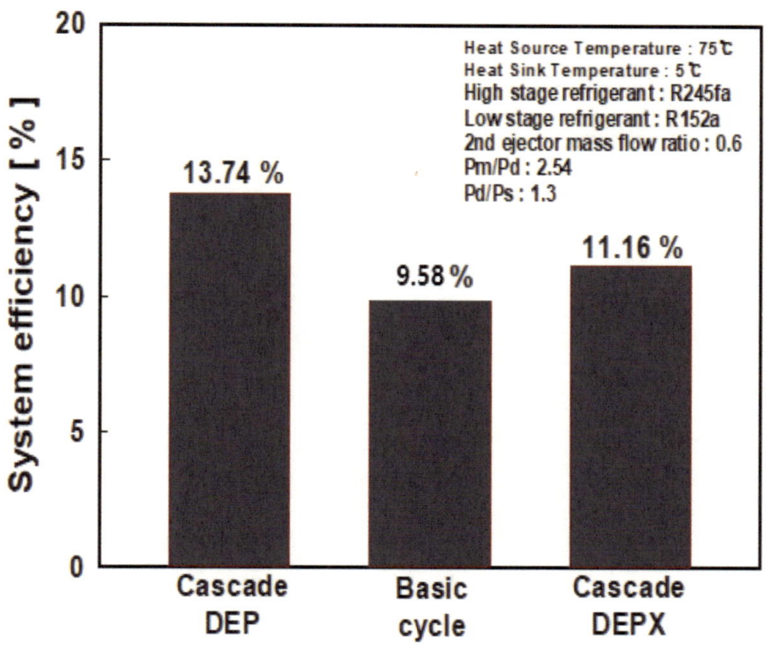

Figure 13.
System efficiency comparison with basic cycle between cascade DEP and cascade DEPX cycles.

lower stage of the cascade system. At the high stage, the DEP and DEPX cycles are applied to improve the system performance. **Figure 13** shows the improved efficiency of the cascade DEP and DEPX cycles over the BOTEC. The high-stage secondary ejector was fixed at a mass flow ratio of 0.6, and the ejector pressure recovery rate was 1.3. The simulation results show that the system efficiency of the cascade DEP and DEPX cycles improved by 43.3 and 16.5%, respectively, over the BOTEC.

3.3 Experiment results

An experimental study was conducted to verify the performance of the ejector under various operating conditions. The operating pressure, discharge pressure, and flow rate were set as the experiment variables as they determine the ejector performance.

Figures 14 and **15** show the device diagram of the ejector performance tester and the actual test device, respectively. The refrigerant side of the experiment apparatus consists of three heat exchangers, a refrigerant pump, two flow meters, an ejector, and flow-control valves. The temperature condition of the operating fluid is achieved by heat exchanging process with the hot water heated by the electric heater at the operating-side heat exchanger (Motive HX), while the pressure is controlled by the refrigerant-side Pump1. The suction fluid is transferred by the ejector traction force from the receiver behind the condenser, and is then vaporized through heat exchanging process with the hot fluid in the heat exchanger (HX). The working fluid discharged from the ejecting part of the ejector in the mixed liquid-gas state passes through the condenser before being condensed into liquid state through heat exchange with low-temperature brine. The mass flow rate of each fluid was adjusted using the installed valves and measured using flow meters

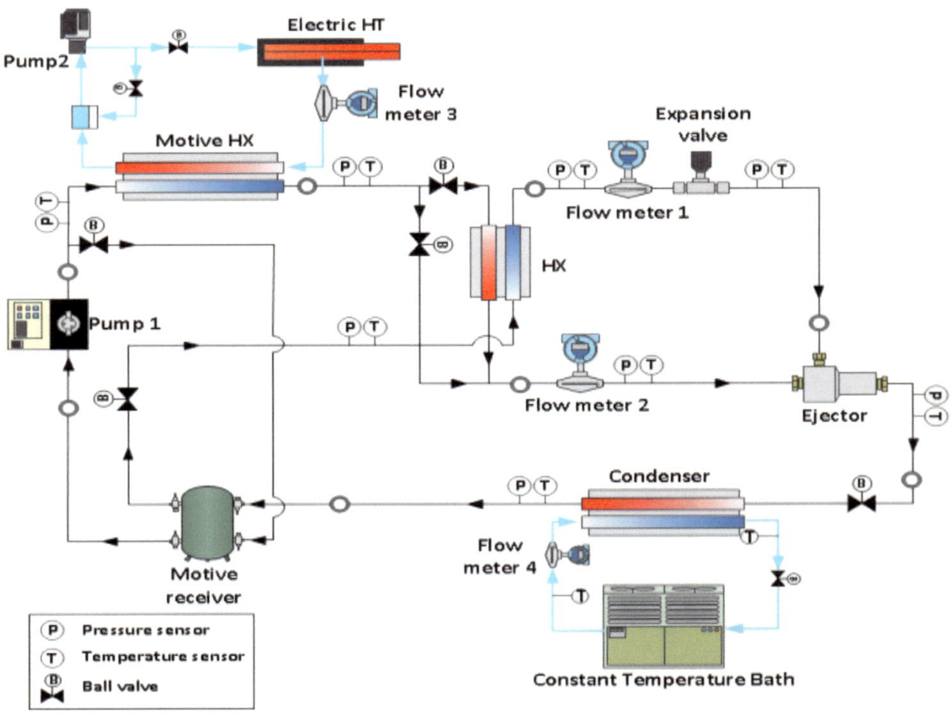

Figure 14.
Schematics of ejector performance experiment.

installed in the operation and suction parts. The basic experiments for ejector performance verification involves

1. design and manufacture of laboratory-scale ejector basic preliminary test equipment

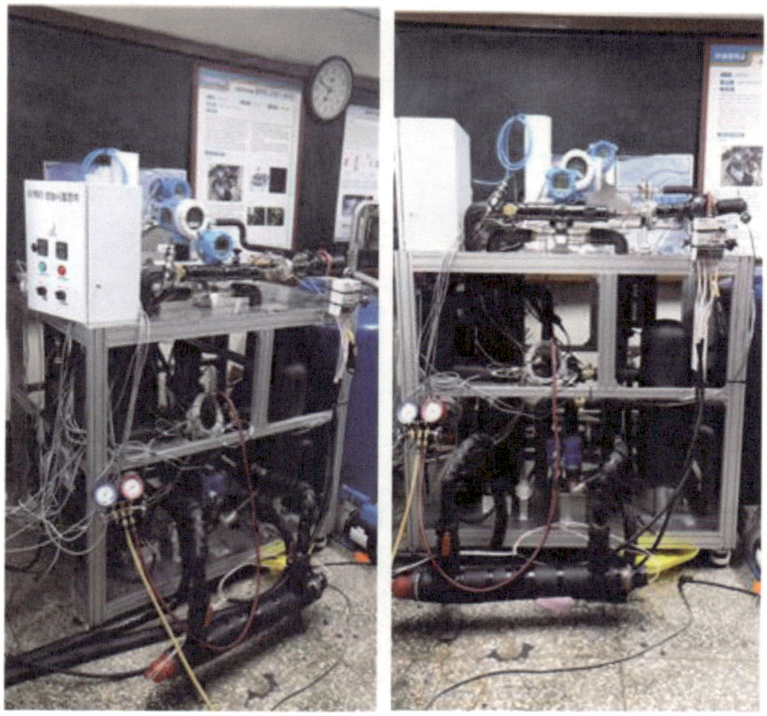

Figure 15.
Experimental apparatus of ejector performance.

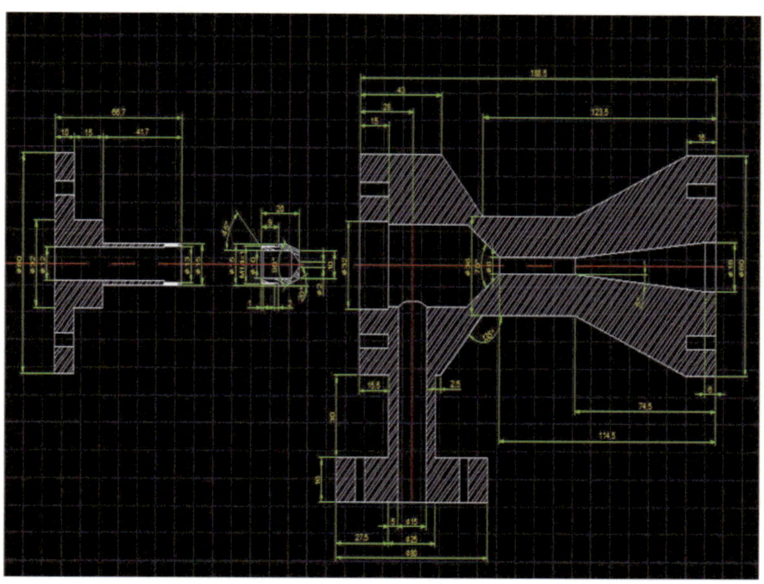

Figure 16.
Design drawing of ejector.

2. performance verification of the ejector according to the working pressure, discharge pressure, and flow rate ratio
3. supply of hot water through the electric heater of the high-temperature part, a heat source, and heat needle supply through a low-temperature thermostat connected to the low-temperature part

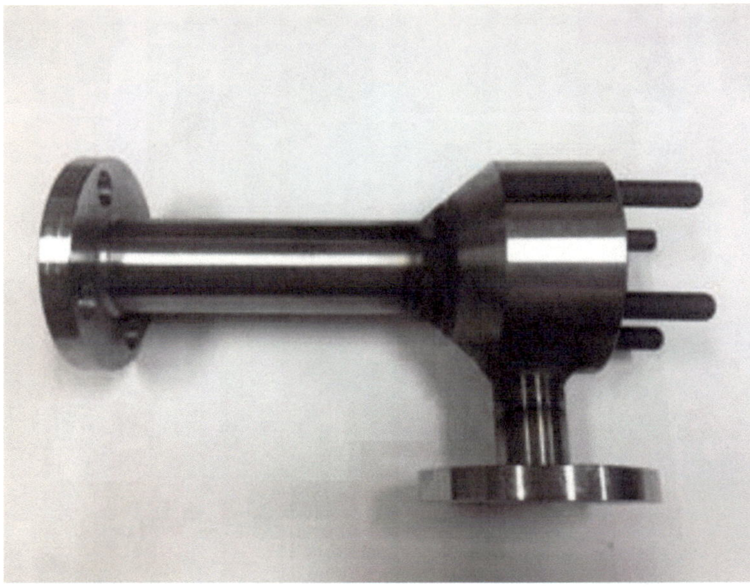

Figure 17.
Manufactured ejector (body).

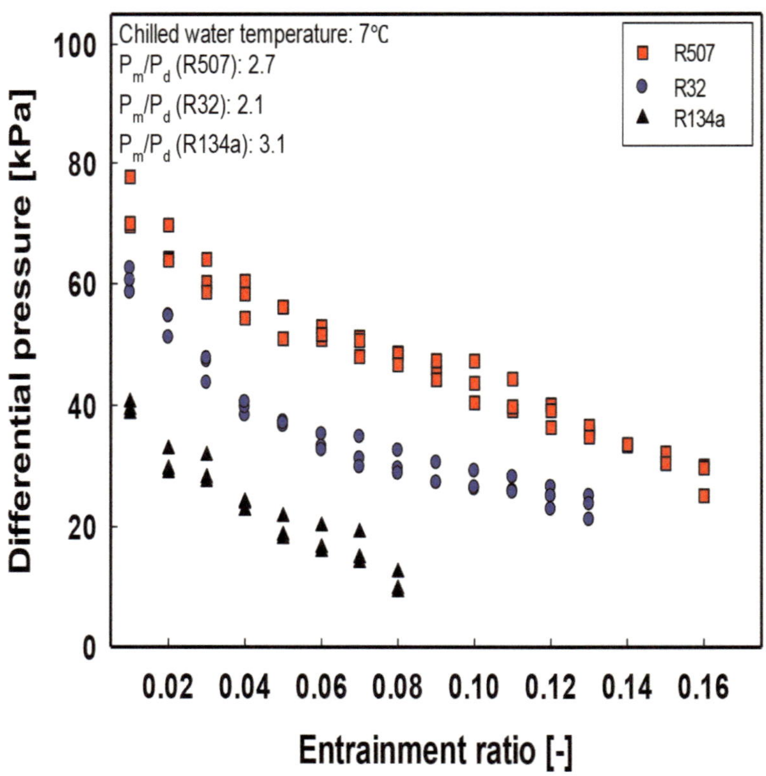

Figure 18.
Differential pressure with respect to entrainment ratio.

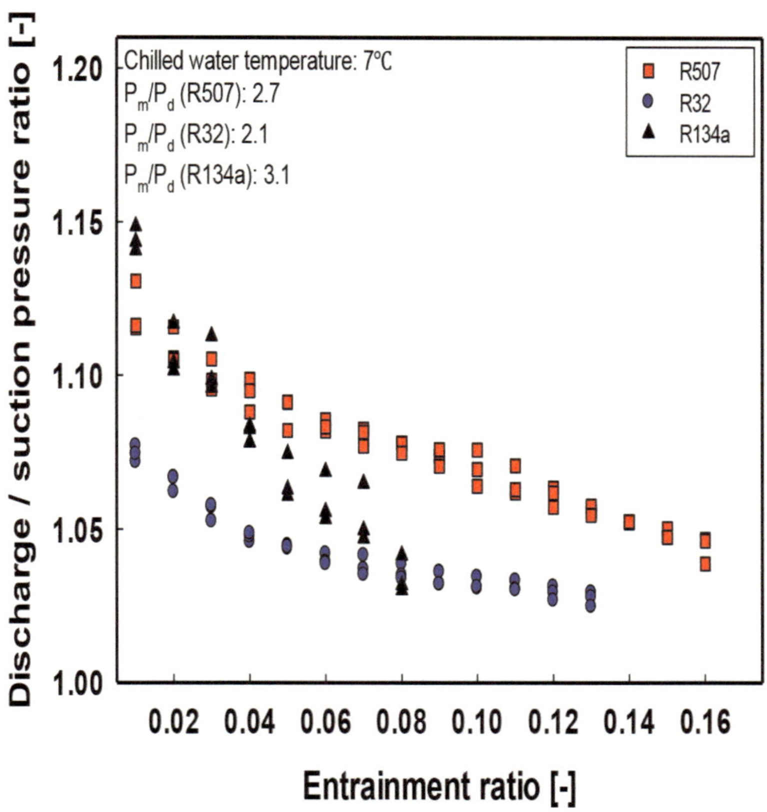

Figure 19.
Discharge/suction pressure with respect to entrainment ratio.

4. selection of working fluid for the refrigeration-combined temperature generator system considering the gas density and efficiency

5. working fluids of R32, R507, and R134a, which have relatively high gas densities, applied as the working fluid at the beginning of the experiment

An ejector was fabricated for the demonstration experiment of the proposed OTEC system, with its specifications and completed product shown in **Figures 16** and **17**, respectively.

Figure 18 shows the difference between the discharge and suction pressures according to the flow rate ratio of the ejector. All the three applied fluids showed a tendency to decrease the pressure difference as the flow rate ratio increased. This was the drop in pressure that occurred as the flow rate below the traction force of the ejector was supplied, and the quantitative level of the pressure difference was different for each working fluid. As R507 has the smallest specific volume among the three working fluids, R507 has the largest pressure difference under the condition that a certain mass flow rate is drawn into the ejector. R32 is a refrigerant with a gas density about 46% less than that of R507 having the gas density of 3.43 g/cm^3, but also with a relatively small pressure differential due to the limited lift of the pump. R134a was the smallest-gas-density refrigerant among the three working fluids, and therefore providing the smallest pressure difference.

Figure 19 shows the ratio of the discharge and suction pressures according to the flow rate ratio of the ejector. This ratio indicates the pressure recovery performance of the ejector. The pressure recovery rate is calculated based on the discharge-unit pressure and the pressure difference between the discharge and suction sides.

On the other hand, the pressure recovery rate of R32 was slightly lower due to its smaller pressure difference and higher discharge pressure, compared to R507.

The experimental study that was conducted using the flow rate ratio of the ejector for the OTEC system is summed up below. As the flow rate ratio of the ejector increased, the pressure difference between the discharge and suction parts decreased, and the level of differential pressure varied with applied refrigerants. As the refrigerant R507 had the smallest specific volume among the three refrigerants that were used, the highest differential pressure was generated in all its flow rate ratio sections. On the other hand, the R134a refrigerant was found to exhibit high-pressure recovery performance (Pd/Ps) although differential pressure is relatively small due to its low discharge pressure.

4. Further plan

Korea Research Institute of Ships and Ocean Engineering aims to complete the OTEC plant at Kiribati by 2021 (**Figure 20**). The production and testing of 1 MW-class OTEC plant such as a 20-kW pilot plant is in progress. After the completion of the domestic demonstration in 2019, onsite construction will be conducted for 2 years. Considering the initial investment cost and the system safety, a closed-temperature generator using R32 refrigerant was decided to be applied. As the current project is to establish a demonstration plant for the first commercial OTEC plant, it will be necessary to apply a high-efficiency system considering economic aspect and performance improvement. The reheating and regeneration cycles mentioned above and the EP-OTEC cycle with ejectors are considered to be competitive in terms of high performance and cost-effectiveness, compared to the conventional Kalina or Uehara cycles.

The following questions should be answered with regard to the future commercialization of the improved OTEC plant:

Figure 20.
3D design of 1-MW OTEC plant in Kiribati (KRISO).

Figure 21.
3D design of advanced future OTEC plant (KRISO).

1. Is the output or efficiency of the improved OTEC plant system excellent enough to be commercialized?

2. Is the application of the improved OTEC system economically competitive?

3. How long does it take to return the initial investment after the first day of operation?

4. Is the improved OTEC system environmentally friendly and safe to operate?

5. Are operations of the improved OTEC system straightforward for educated operators?

6. Can the improved OTEC system be easily automatized and be under the unmanned operation?

Through the study of improved OTEC based on the above considerations, we expect that high-efficiency stabilized power plants, shown in **Figure 21**, become available in the near future, while the first commercializable OTEC will be running at Kiribati in a few years.

Acknowledgements

This work was financially supported by the National R&D project of "Development of 1MW OTEC demonstration plant (4/6)" (PMS4080) funded by the Ministry of Oceans and Fisheries of the Republic of Korea.

Author details

Hosaeng Lee[1*], Seungtaek Lim[1], Jungin Yoon[2] and Hyeonju Kim[3]

1 Seawater Energy Plant Research Center, Korea Research Institute of Ships and Ocean Engineering, Goseong-gun, South Korea

2 Department of Refrigeration and Air-Conditioning Engineering, College of Engineering, Pukyong National University, Busan, South Korea

3 Offshore Plant and Marine Energy Research Division, Korea Research Institute of Ships and Ocean Engineering, Daejeon, South Korea

*Address all correspondence to: hslee@kriso.re.kr

References

[1] Avery WH, Wu C. Renewable Energy from the Ocean: A Guide to OTEC. England: Oxford University Press; 1994

[2] Galbraith K. Generating Energy from the Deep. The New York Times; 29 Apr 2009

[3] Uehara H, Nakaoka T. Development and Prospective of Ocean Thermal Energy Conversion and Spray Flash Evaporator Desalination. Saga, Japan: Saga University; 2007

[4] Finney AK. Ocean thermal energy conversion. Guelph Engineering Journal. 2008;(1):17-23

[5] Salz K. An assessment of the performance and potential of OTEC innovation clusters worldwide [thesis]. Delft University of Technology; 2018

[6] Uehara Cycle in Saga Univ [Internet]. Available from: http://www.ioes.saga-u.ac.jp/en/facilities/ioes_facilities

[7] Amyra MY, Othman N'a, Sarip S, Ikegami Y, Chik MAT, Othman N, et al. Simulation study on enhancing hydrogen production in an ocean thermal energy (OTEC) system utilizing a solar collector. Journal Teknologi (Sciences & Engineering). 2015;**77**(1):23-31. DOI: 10.11113/jt.v77.4145

[8] Aydin H. Performance analysis of a closed-cycle ocean thermal energy conversion system with solar preheating and superheating [thesis]. University of Rhode Island; 2013

[9] Lee HS, Lim ST, Moon JH, Kim HJ. Performance assessment for high Temperature OTEC plant. Proceedings of Renew 2016, 2nd International Conference on Renewable Energies Offshore. 2016:475-480

[10] Lee HS, Kim HJ, Jung DH, Moon DS. A study on the improvement for cycle efficiency of closed-type OTEC. Journal of the Korean Society of Marine Engineering. 2011;**35**(1):46-52. DOI: 10.5916/jkosme.2011.35.1.046

[11] Yoon JI, Seol SH, Son CH, Jung SH, Kim YB, Lee HS, et al. Analysis of the high-efficiency EP-OTEC cycle using R152a. Renewable Energy. 2017;**105**:366-373. DOI: 10.1016/j.renene.2016.12.019

Chapter 5

Analysis and Development of Closed Cycle OTEC System

Estela Cerezo Acevedo, Jessica G. Tobal Cupul,
Victor M. Romero Medina, Elda Gomez Barragan
and Miguel Angel Alatorre Mendieta

Abstract

In this chapter, we present the methodology for the selection of the working fluid, the environmental and working conditions for operation, and the development carried out for the design of a closed cycle OTEC prototype plant. This prototype uses the temperature difference between the cooler deep waters and the warmer surface waters of the Mexican Caribbean Sea to feed a thermal machine capable of generating 1 kW of electrical energy; and it works with an organic Rankine cycle, composed of a pump, a turbine, and two heat exchangers. The advances carried out in installing the prototype are also presented.

Keywords: OTEC, working fluid, organic Rankine cycle, Mexican Caribbean Sea

1. Introduction

This chapter shows the design and progress of the installation of the first 1 kWe OTEC prototype plant carried out in Mexico to be tested first at the laboratory level and then at potential thermal gradient sites in Mexico. Mexican Caribbean is a potential site because of its physical characteristics being a renewable energy deposit and a resource in Mexico that is located in an area of 98,000 km^2, with 825 km of littoral, corresponding to its exclusive economic zone (EEZ), adjoining the sea portions of the Republic of Cuba, Republic of Honduras, and Belize [1]. In the Caribbean Sea, as the surface temperature is very stable and the depth of 1000 m is not reached far from the coast, there are potential areas to install an OTEC plant in Cozumel Island, Punta Allen, Tulum, Sian Ka'an, Xcalak, Mahahual, and Chinchorro Bank [2]. It should be noted that all these areas meet the thermal gradients greater than or equal to 20°C and are located at 700 m depth and at a distance from the coast less than or equal to 10 km [3], as shown in **Figure 1**.

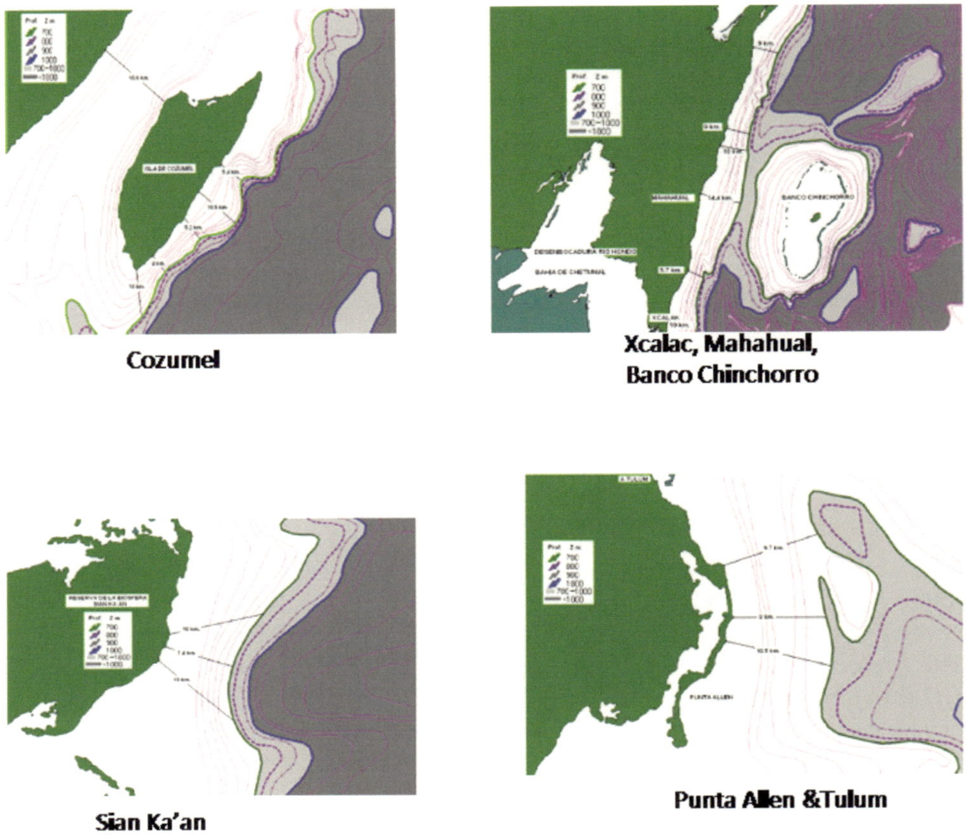

Figure 1.
Potential sites in the Mexican Caribbean Sea. Source: Barcenas (2014).

2. Operation temperature conditions

To install an OTEC plant, its proximity to the land and a city must be considered.

2.1 Thermal potential of the Mexican Caribbean

In 2014, Barcenas mentions that the technically available power in the Mexican Caribbean is 2000 MWe, taking into account the potential OTEC locations shown in **Figure 1** with a thermal gradient of 20°C [2].

Analyzing the results of [2], we observe that the site in which an OTEC plant could be installed taking into account its proximity to land and a city is the island of Cozumel as the isobath of 700 m, closer to the coast, is located 4 km from the southeast coast of this island.

Another place where the installation of an OTEC plant would be very useful is in Punta Allen since it is a town that lacks electricity, drinking water, and sewage system, although the nearest isobath of 700 m is 9 km away from the coast.

2.2 Caribbean Sea temperature profiles

The key parameter is the temperature at different depths. In **Figure 2**, the annual average of the temperature gradient is shown. A minimum gradient in March (18.38°C) and a maximum in October (22.21°C) are observed.

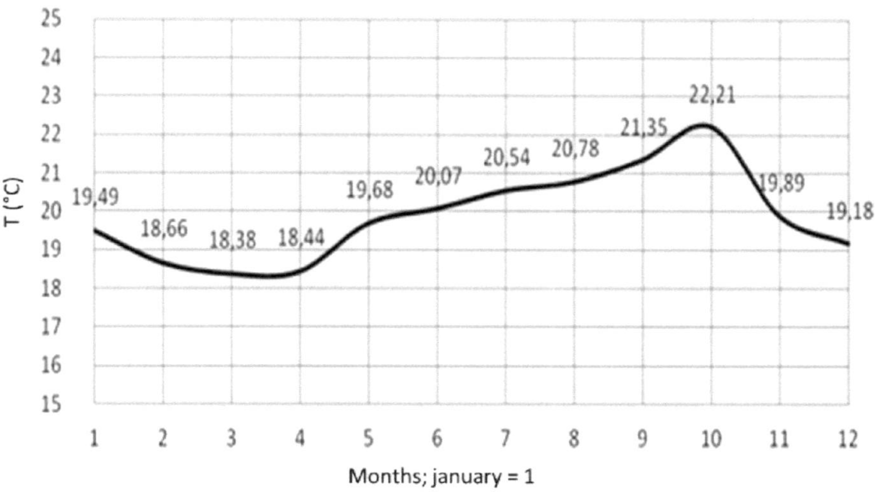

Figure 2.
Difference in temperature average between 0 and 1000 m. Source: Barcenas (2014).

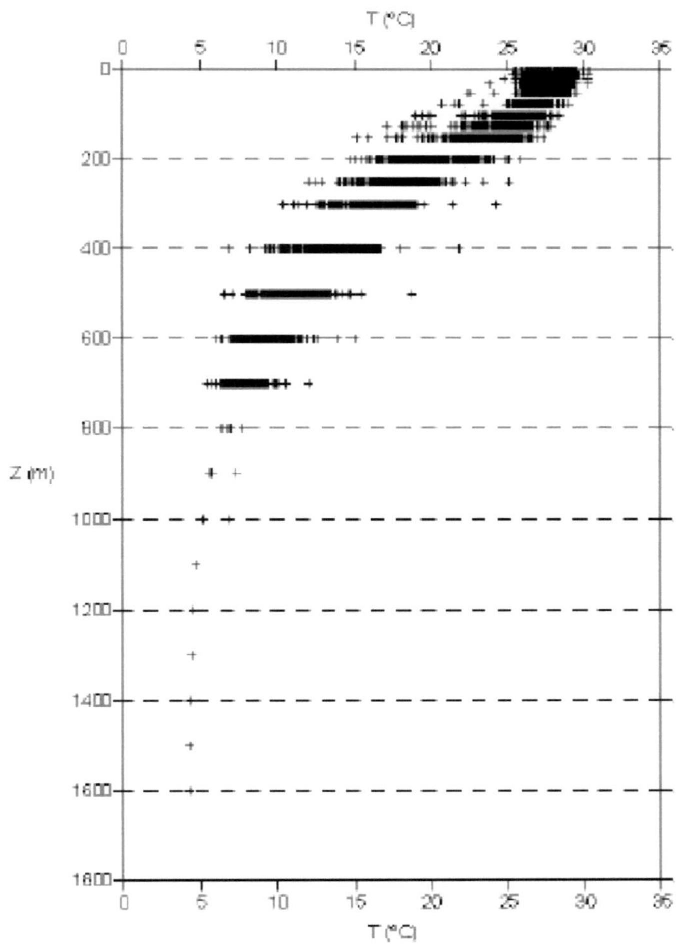

Figure 3.
Dispersion of Caribbean Sea temperatures with depth (latitudes 22°N to 17°N and −88.5°W and −84.5°W). Source: Barcenas (2014).

Barcenas (2014), based on the data collected to date by the National Oceanic and Atmospheric Administration, dispersed the temperature with respect to the depth for the Caribbean area (latitudes 22°N and 17°N and longitudes −88.5°W to −84.5°W). In **Figure 3**, we can observe this dispersion.

In the same reference, it is mentioned that the annual average is for the minimum temperature at 700 m of 7.69°C and the maximum surface is 27.46°C. Therefore, our initial temperature conditions will be minimum temperature 7°C and maximum temperature 27°C to overestimate.

3. Working fluid selection

The selection of the working fluid is of vital importance to achieve the maximum efficiency of the cycle in the OTEC plant, since the thermodynamic performance of the fluid has an impact on the size of the components of the plant and finally on the cost of the same [4].

In the following sections is described the methodology for selecting the working fluid in the OTEC-CC prototype plant that is being built at the University of the Caribbean according to different selection criteria.

3.1 Selection criteria

The working fluid is selected from commercial version of EES according to next criteria: (a) turbine and exchangers dimensions (φ and β); (b) environmental impact, ozone destruction potential (ODP) and global warming potential (GWP); and (c) safety (toxicity and flammability) and others (saturation pressure at 15.56° C and costs).

φ and β are defined as follows:

$$\varphi = \left(\frac{k^3\rho^2 h_{lg}}{\mu}\right)^{1/4} \tag{1}$$

$$\beta = P_v \Delta h_{lg} M \tag{2}$$

where P_v is the saturation pressure at 15.59°C (kPa), h_{lg} is the evaporation enthalpy (kJ/kg), M is the molecular weight (kmol/kg), k is the liquid fluid thermal conductivity (W/m*K), ρ is the liquid density (kg/m^3), and μ is the liquid dynamic viscosity (Pa*s).

φ is associated to the turbine diameter: if there is a maximum diameter, there is a minimum heat loss transfer to atmosphere. β is inversely proportional to mass flow, so the greater value of β can be found elsewhere [4, 5].

ODP is referred to ozone destruction capacity on stratosphere according to R11 ozone destruction capacity (which ODP = 1). GWP measures global warming potential according to carbon dioxide for 100 years [6]. For toxicity and flammability parameters, the occupational exposure limit (OEL) and the lower flammable limit (LFL) were investigated. OEL indicates the maximum amount that a person can be exposed for hours per week: a high value means low toxicity. LFL indicates the minimum amount of fluid concentration to be flammable in contact with the air: a high number means low flammability [6].

P_v was calculated on EES at 15.56°C, because 15.56°C is the evaporator's estimated phase change temperature. P_v relevance is because of the pressure effect cycle total efficiency; it could be required to attach a vacuum system or increase pump pressure [7].

The working fluid cost is in American Dollar. This represents an average fluid cost when there was more than one provider.

3.2 First selection of working fluid

ODP, GWP, and P_v are the parameters used as the first filter of the 50 fluids available in the EES program. On this phase those fluids that had any of the following characteristics were dismissed: (1) ODP > 0; (2) GWP > 2500 (according to European legislation, it is the same limit used for fixed refrigeration systems) [8]; and (3) P_v below atmospheric pressure (101.35 kPa) and greater than 1 MPa.

According to the first limit for P_v, if there would be a P_v lower than the atmospheric pressure, it would be necessary to make a vacuum to achieve fluid evaporation [9]. The second limit for P_v was determined to avoid creating a high pressure with the pump.

From this fist selection of working fluids, the remaining were ammonia, isobutane, n-butane, propane, propylene, R134a, R152a, R407C, and R600a. These fluids were used for the next evaluations.

3.3 Evaluations

The evaluation of these nine remaining fluids was made using six parameters: GWP, OEL, LFL, φ, β, and the cost, as listed in **Table 1**. These were normalized in order to make a comparison with each other.

Each parameter has a different weight according to the importance that the parameter represents in the evaluation, and therefore there is a different weighting in each of them.

In this study, five weightings with specific weights for each parameter were considered, and each weighting corresponds to a specific work fluid selection objective.

3.4 Weightings and results

3.4.1 Weighting 1

Environmental impact and the cost of the fluid were considered the most important parameters, so 50% of the evaluation weight was divided equally for these parameters. Thirty-four percent of the evaluation weight was divided equally

Fluid	GWP	LFL (%)	OEL (PPMv)	φ	B	Cost($/kg)
Ammonia	1	16.7	25	10910.54	15224309.1	0.63
Isobutane	20	1.6	1000	18849.26	5193836.84	14.22
n-butane	20	2	1000	19411.61	3868581.12	0.06
Propane	20	2.1	1000	16321.13	11505975.77	18.59
Propylene	20	2.7	500	16605.01	13450526.52	2.49
R134a	1370	0	1000	34352.59	9443644.5	9.01
R152a	133	4.8	1000	25008.43	8558070.76	3.97
R407c	1700	0	1000	31875.68	12855787.83	10.44
R600a	20	1.6	1000	18612.92	5126886.38	14.22

Table 1.
Evaluated fluids.

between the parameters related to safety. The remaining 16% was divided among the other parameters (**Table 2**).

The results of this weighting are presented in **Table 3**, where it is shown that n-butane was the best fluid for this weighting with 69.69% effectiveness, followed by R152a with 68.45% and R134a with 63.71%.

3.4.2 Weighting 2

In this weighting scheme, the construction of an OTEC plant at the laboratory level was the main scenario to determine the weight of each parameter in the evaluation. Therefore 50% was directed to parameters related to safety, assigning 30% to toxicity and flammability with 20%. Then, 30% of the evaluation weight was divided equally between the environmental impact and the fluid cost. The remaining 20% was divided equally between the parameters referring to the size of the components of the OTEC plant (see **Table 4**).

Factor	Weight
GWP	0.25
B	0.08
Φ	0.08
Costs	0.25
Toxicity	0.17
Flammability	0.17

Table 2.
Weighting 1.

Fluid	Points (%)	Fluid	Points (%)
n-butane	69.69	R407C	58.48
R152a	68.45	Isobutane	51.26
R134a	63.71	R600a	51.13
Propylene	63.6	Propane	49.03
Ammoniac	59.84		

Table 3.
Results of weighting 1.

Factor	Weight
GWP	0.15
B	0.10
Φ	0.10
Costs	0.15
Toxicity	0.30
Flammability	0.20

Table 4.
Weighting 2.

In the **Table 5**, the results of this weighting are observed, with R134a being the fluid with the highest percentage, followed by R407C and R152a with 73.45 and 66.46%, respectively.

3.4.3 Weighting 3

The conviction that safety conditions could be controlled by taking appropriate measures during the OTEC prototype plant tests was the scenario considered for weighting 3 (see **Table 6**). Therefore, the safety parameters were considered 50% less important than the other parameters.

As a result of this weighting (see **Table 7**), R152a, R134a, and propylene fluids were the refrigerants that obtained the highest percentage, with respective values of 64.84, 64.04, and 63.87%.

Fluid	Points (%)	Fluid	Points (%)
R134a	75.58	Propylene	53.57
R407C	73.45	Isobutane	52.92
R152a	66.46	R600a	52.76
n-butane	63.54	Ammonia	42.61
Propane	53.97		

Table 5.
Results of weighting 2.

GWP	0.20
B	0.20
Φ	0.20
Costs	0.20
Toxicity	0.10
Flammability	0.10

Table 6.
Weighting 3.

Fluid	Points (%)	Fluid	Points (%)
R152a	64.84	n-butane	57.07
R134a	64.04	Propane	42.89
Propylene	63.87	Isobutane	43.60
R407C	62.51	R600a	43.28
Ammonia	60.92		

Table 7.
Results of weighting 3.

3.4.4 Weighting 4

Considering all the equally important parameters was the objective of the weighting (see **Table 8**). Therefore, 100% of the evaluation weight was divided equally among the six parameters.

The results of **Table 9** showed R134a (with 70.04%), R407C (68.76%), and R152a (62.64%) as the fluids that best adapted to the weighting.

3.4.5 Weighting 5

For weighting 5 (of **Table 10**), the parameters referring to the OTEC plant equipment were considered the most important, so that 50% of the weight was divided equally into the value of φ and β. Then, another 40% was divided equally into the parameters related to environmental impact and safety; the latter has 10% for toxicity and flammability. Finally, the remaining 10% was allocated to the cost of the fluid.

Factor	Weight
GWP	0.16
B	0.16
Φ	0.16
Costs	0.16
Toxicity	0.16
Flammability	0.16

Table 8.
Weighting 4.

Fluid	Points (%)	Fluid	Points (%)
R152a	64.84	n-butane	57.07
R134a	64.04	Propane	42.89
Propylene	63.87	Isobutane	43.60
R407C	62.51	R600a	43.28
Ammonia	60.92		

Table 9.
Results of the weighting 4.

Factor	Weight
GWP	0.20
B	0.25
Φ	0.25
Costs	0.10
Toxicity	0.10
Flammability	0.10

Table 10.
Weighting 5.

Fluid	Points (%)	Fluid	Points (%)
R407C	66.54	Propane	52.41
R134a	66.33	n-butane	48.88
R152a	62.02	Isobutane	43.52
Propylene	60.62	R600a	43.12
Ammonia	56.23		

Table 11.
Results of weighting 5.

Position	Weighting 1	Weighting 2	Weighting 3	Weighting 4	Weighting 5
1	n-butane	R134a	R152a	R134a	R407C
2	R152a	R407C	R134a	R407C	R134a
3	R134a	R152a	Propylene	R152a	R152a

Table 12.
Frequency of the first three positions in evaluations.

The fluids R407C, R134a, and R152a were the ones that best suited the weighting, according to the evaluation results (see **Table 11**).

3.5 Selected fluid

The summary of evaluations is given in **Table 12**, where it is shown that R134a, R152a, and R407C were the fluids that had the highest frequency along the different weightings. The R134a appeared in one of the first three positions in all evaluations and obtained the first position in 2/5 evaluations, the second position in 2/5 evaluations, and the third position in 1/5 evaluations. R152a also appeared in one of the first three positions in all the evaluations and obtained the first position in 1/5 evaluations, the second position in 1/5 evaluations, and the third position in 3/5 evaluations. Finally, the R407C did not appear in 2/5 evaluations; however, it obtained the first position in one evaluation and the second position in the two remaining evaluations.

Finally, R152a (1,1-difluoroethane, $C_2F_2H_4$) was selected for the OTEC-CC prototype plant, since the laboratory conditions will be controlled throughout the process. The working fluid was evaluated under ideal Rankine cycle with temperatures of hot and cold source of 27 and 7°C, respectively, obtaining the thermal efficiency (η_T) of 3.445%.

4. A 1 kWe CC-OTEC prototype plant

The main components in this prototype are the evaporator, condenser, turbine, and pump; a steam separator and a tank are also included. For the prototype to be tested in the laboratory, heating and cooling systems are included, as shown in **Figure 4**. All the components relate to each other using pipes and valves. R-152a at compressed liquid state is transported by the pump to the evaporator, where the working fluid is evaporated into a saturated vapor state by warm water from the heating system. The vapor of the working fluid passes through the steam separator to prevent liquid to enter to the turbine, and it could damage the blades. Then the

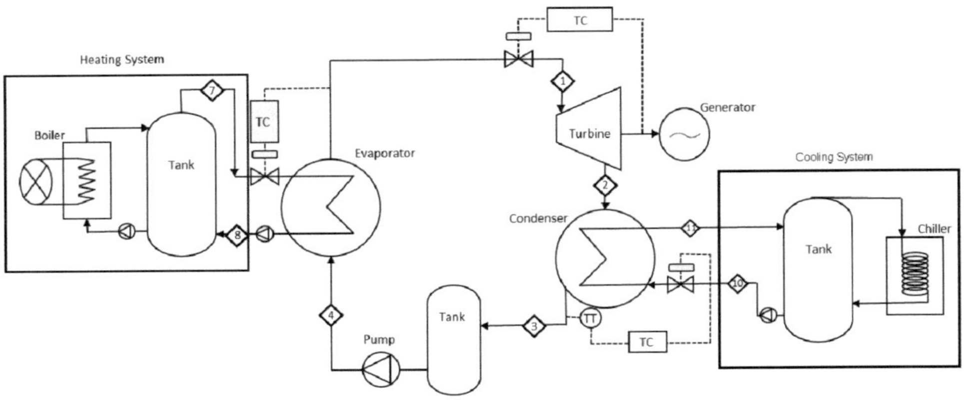

Figure 4.
Closed cycle OTEC prototype plant.

vapor drives the turbine and the connected electrical generator to produce 1 kW of electricity. The mixture vapor from the turbine is cooled into a saturated liquid state by the cool water in the condenser, and then the saturated liquid of the working fluid is transported by the pump to the evaporator to close the cycle. However, because of the small temperature difference between the warm water and cool water, the thermal efficiency of prototype OTEC plant is very low, about 3%. The balances of mass and energy were made for the diagram of **Figure 4**.

The heat added in kJ/kg (q_e) is

$$q_e = h_1 - h_4 \tag{3}$$

where $\dot{m}$ is mass flow and h is the enthalpy at the indicated state point.

The turbine's real work in kJ/kg (w_{Tr}) is

$$w_{Tr} = h_1 - h_{2r} \tag{4}$$

where

$$h_{2r} = h_1 - \eta_T(h_1 - h_{2s}) \tag{5}$$

and η_T is the turbine efficiency, h_{3r} is the real enthalpy, and h_{3s} is the isentropic enthalpy.

The heat rejected in kJ/kg (q_c) is

$$q_c = h_2 - h_3 \tag{6}$$

and the pump work is

$$w_p = h_4 - h_3 \tag{7}$$

4.1 Carnot efficiency

It is the maximum efficiency that this system can have that will operate between two thermal energy deposits at temperatures T_L = 280.2 K and T_H = 300.3 K.

$$\eta_{carnot} = \left(1 - \frac{T_L}{T_H}\right)100 \tag{8}$$

Parameter	
η_{Carnot}	6.7%
η_{th}	2.4%
X	0.99
Q_e	45.31 kW
Q_c	44.20 kW
Wp	0.017 kW
Wt	1.11 kW

Table 13.
Heat and mass balance for 1 kWe prototype plant.

Parameter	Unit	R152a				Water			
Lines		1	2	3	4	7	8	9	10
Temperature	°C	22.5	22.5	12	11.61	27	23	7	10
Pressure	MPa	0.55	0.55	0.4	0.39	0.13	0.1	0.11	0.1
Mass flow	kg/s	0.15	0.15	0.15	0.15	2.71	2.71	3.51	3.51

Table 14.
Flow mass results.

The real efficiency of this system is

$$\eta_{th} = 1 - \frac{q_c}{q_e} \tag{9}$$

The results of the heat and mass balance are presented in **Table 13**.

The thermodynamic evaluations were carried out with the EES program.

The results of mass flow of the working fluids and intake seawater flow for the evaporator and condenser are presented in **Table 14**. 2.71 kg/s of surface seawater at 27°C and 3.51 kg/s of deep seawater at 7°C are necessary in this prototype.

5. Advances of the installation of the OTEC prototype plant

This section describes the progress of the installation of the OTEC prototype plant to generate 1 kWe, which will be tested in the laboratory and later the sites of interest in Mexico.

5.1 Pipe sizing

Once the mass and energy balances were made, the sizing of the pipe diameters (ϕ_t) was carried out based on the mass flow equation (Eq. 12), which is a physical quantity that expresses the variation of mass with respect to time in a specific area (A_T).

$$\dot{m} = \rho v A_T \tag{10}$$

where

$$A_T = \frac{\pi {\phi_t}^2}{4} \tag{11}$$

and

$$\phi_t = \sqrt{\frac{4\dot{m}}{\rho v \pi}} \tag{12}$$

In all cases, the calculation for pipe diameter size was made in mm, and subsequently the result was adjusted to the immediate higher pipe diameter size for commercial pipes. For example, if $\phi_t = 210$ mm, around 0.82 in, a commercial pipe of 1 in of diameter was selected.

In the case of the pipe for the liquid refrigerant, a speed of 1.5 m/s was considered. For the refrigerant in gaseous state a speed of 20 m/s (**Table 15**).

5.2 Installation

The installation of the OTEC prototype plant was carried out based on **Figure 4**. The storage tank (1100 L) of the cooling system was connected with the plate condenser (Line 10 and 11 of **Figure 4**) by a PVC pipeline Schedule 40 with a nominal diameter of 1½″.

To connect the storage tank of the heating system (1100 L) with the plate evaporator and a ¾ HP surface pump (lines 7 and 8, **Figure 4**), PVC pipeline 40 of 1½″ and heavy duty 2 × 12 cable were used. Rheem 89V40 electric heater was connected to storage tank with PVC pipeline 40 of ¾″ (hydraulic line) and to a 220 V socket with heavy duty 3 × 12 cable (electrical line).

To complete the installation of the OTEC prototype plant (**Figure 5**), the turbine and the working fluid pump must be connected. The turbine was designed because

Dates	R152a						Water			
Line	**1**	**2**	**3**	**4**	**5**	**6**	**7**	**8**	**9**	**10**
Diameter (in)	°C	1	1	1	0.5	0.5	1.5	1.5	1.5	1.5
Materials	Cu						PVC			

Table 15.
Pipe diameter and materials.

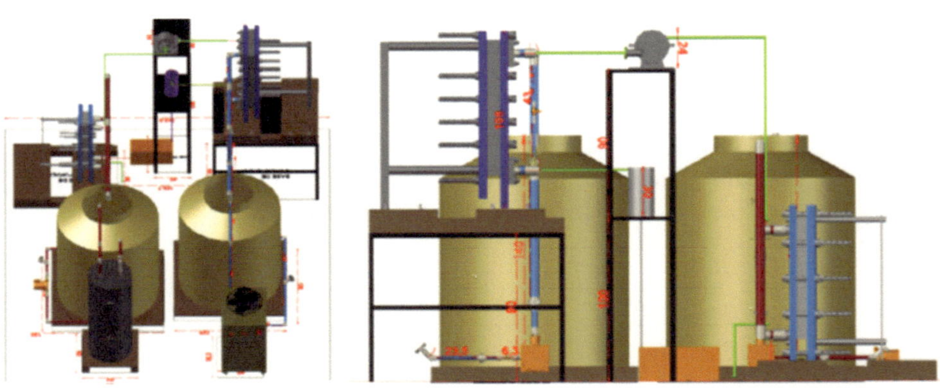

Figure 5.
3D prototype installation diagram.

there is no commercial turbine with the characteristics of the project and at the moment it is in manufacturing process.

6. Conclusions

According to the temperature profile of the Mexican Caribbean Sea, it was determined that the temperatures for the operating conditions are 27 and 7°C for surface seawater and 700 m depth, respectively. A closed cycle OTEC system was also selected for the expected dimensions of the turbine-electric generator, and according to the assigned weights, the best working fluid was R-152a, being compatible with copper, pipe material, and equipment that are planned to be used.

As expected, the maximum efficiency is low (η = 6.7%) since the difference in temperature between the heat source and the heat sink is small (ΔT = 20°C); therefore, the effective thermal efficiency will be much lower (η = 2.4%) than the other plants that use the Rankine cycle, such as thermoelectric plants; however, it should be noted that fossil fuels are not consumed and the source of heat is free and inexhaustible. Likewise, it was observed that only 2% of the power produced by the generator would be used to power the pump.

Up to now, the heating and cooling systems have been installed; we continue working on the installation and testing of the OTEC prototype plant.

Acknowledgements

This work was financially supported by the Mexican Centre of Innovation in Ocean Energy supported by the National Council for Science and Technology and the Mexican Secretariat of Energy.

Author details

Estela Cerezo Acevedo[1*], Jessica G. Tobal Cupul[1], Victor M. Romero Medina[1], Elda Gomez Barragan[2] and Miguel Angel Alatorre Mendieta[3]

1 Department of Basic Sciences and Engineering, University of Caribe, Cancun, Quintana Roo, Mexico

2 Academy of Environmental Engineering, Mountain Technological Institute, Tlapa de Comonfort, Guerrero, Mexico

3 Institute of Marine Sciences and Limnology, National Autonomous University of Mexico, Mexico City, Mexico

*Address all correspondence to: ecerezo@ucaribe.edu.mx

References

[1] Bárcenas J. El mar Caribe de México y su potencial energético renovable. Cuba: Memorias IX Congreso Internacional de Ciencias del; 2012

[2] Bárcenas J. Evaluación del potencial energético renovable del Caribe Mexicano. México, D.F., México: Instituto de Ciencias del Mar y Limnología; 2014

[3] Garduño E et al. Conversión de Energía Térmica Oceánica (OTEC) Estado del Arte. Cemie-Océano: Universidad Autónoma de Campeche; 2017

[4] Yoon J-I et al. Efficiency comparison of subcritical OTEC power cycle using various working fluids. Heat and Mass Transfer. 2014;**50**:985-996

[5] Avery WH, Wu C. Renewable Energy from the Ocean: A Guide to OTEC. New York, Oxford: Oxford University Press; 1994

[6] Bernal M. Estudio técnico y económico de una planta OTEC y sus usos secundarios en México. (Tesis de maestría). Ciudad de México: Universidad Nacional Autónoma de México; 2016

[7] Calm JM, Hourahan GC. Physical, safety, and environmental data for current and alternative refrigerants. In: Proceedings of 23rd International Congress of Refrigeration (ICR2011); Prague, Czech Republic: s.n.; August 2011. pp. 21-26

[8] Domanski PA. Evolution of refrigerant application. International Congress of Refrigeration. 1999;**4**:131

[9] REGLAMENTO (UE) N° 517/2014 del Parlamento Europeo y del Consejo, de 16 de abril de 2014, sobre los gases fluorados de efecto invernadero y por el que se deroga el Reglamento (CE) N° 842/2006. Unión Europea. s.l.: Diario Oficial de la Unión Europea

Section 3

Case Studies and Assessments

Chapter 6

Research and Development Activities of Ocean Thermal Energy-Driven Development in Malaysia

A. Bakar Jaafar, Mohd Khairi Abu Husain and Azrin Ariffin

Abstract

The search for potential investors in the conversion of ocean thermal energy to power or hydrogen, and its spinoff projects in Malaysia and the region, continues. In the meantime, several pre-feasibility studies have been completed for selected sites, including that of Pulau Layang-Layang and Pulau Kalumpang (Sabah, Malaysia); Timor-Leste, and off Pulau Weh (Aceh, Indonesia). Various research projects have been completed such as the conversion of solar-thermal to the chilled-water system; the cooling of tropical soils for the culture of temperate crops; the design of offshore structure off the continental slope; hydrogen fuel production and distribution, deep seawater properties to reduce obesity, cholesterol and blood pressure; and the legal-institutional framework for the development of ocean thermal energy conversion. UTM Ocean Thermal Energy Centre (UTM OTEC) has entered into the Collaborative Research Agreement with the Institute of Ocean Energy of Saga University (Japan) to undertake joint research for the development of an experimental rig that introduces a hybrid system with stainless steel heat exchanger. Other aspects of this joint research would include a new design for 3 kW turbine, the introduction of nano-working fluids, the eDNA of intake waters, and improved productivity in the culture of high-value marine produce and products.

Keywords: renewable energy, ocean thermal energy, OTEC, Malaysia, sustainable development

1. Introduction

1.1 Establishment of UTM Ocean Thermal Energy Centre (UTM OTEC)

UTM OTEC was established on 3 January 2013, and Emeritus Prof Dr. Md Nor Musa was its first Director, and now, the Director is Prof Dato' Ir Dr. A. Bakar Jaafar, FASc. The main purpose of this Centre is as much to promote investment in the commercialisation of the ocean thermal energy conversion technology, as to undertake its further development, research, innovation, and entrepreneurship.

1.2 Founding co-chair/director

Prof Dato'Ir Dr. A. Bakar joined Universiti Teknologi Malaysia, first joined the university as a Visiting Professor, UTM Perdana School of Science, Technology, and Innovation on 1 June 2012, and later, became a full Professor on 1 August 2013. Prior to this, he was nominated and elected as one of the 21 Members of the UNCLOS Commission on the Limits of Continental Shelf [https://www.un.org/Depts/los/clcs_new/commission_members.htm#Members] over the three 5-year terms: 1997–2002, 2002–2007, and 2007–2012. He also served as an Adviser to the National Continental Shelf Committee, Secretariat to the National Security Council, Malaysia over that period, and responsible inter alia the Malaysian Marine Survey in the southern part of the South China Sea (2006–2008) (MyMRS).

1.3 Malaysian Marine Survey (MyMRS)

One of the consequential findings of the Survey was the temperature profiling of the deep waters of the States of Sabah and Sarawak, where the temperature drops at the surface from 27° to 4°C at a depth greater than 1200 m. Thus, Malaysia has since established its potential to extract the heat from the waters and convert it into power by deploying the fast emerging and most impactful ocean thermal energy conversion technology [https://issuu.com/asmpub/docs/eset_study_report, pages 92–93].

2. Science and technology research partnership for sustainable development (SATREPS)

The latest development is that, through the Centre, UTM has entered into Agreement with Saga University on 27 March 2019 to undertake joint 'Development of Advanced Ocean Thermal Energy Conversion Technology for Low Carbon Society and Sustainable Energy System: First Experimental OTEC Plant in Malaysia' under the JST-JICA Science and Technology Research Partnership for Sustainable Development (SATREPS) Programme (https://www.jst.go.jp/global/english/about.html) over the five-year period, 2019–2024. This Project is led by UTM Ocean Thermal Energy Centre (UTM OTEC) under the leadership of its Director and supported by 10 sub-project leaders from five universities: Universiti Kebangsaan Malaysia (UKM), University of Malaya (UM), Universiti Malaysia Terengganu (UMT), Universiti Putra Malaysia (UPM), and Universiti Teknologi Malaysia (UTM) [http://utm.my/satreps].

3. UTM Ocean Thermal Energy Centre (UTM OTEC) activities

3.1 Pre-2013

The idea of 'ocean thermal energy' was cited by the founder in his internship report (**Figure 1**) as part and parcel of fulfilling an academic requirement of Master degree in Environmental Science, Miami University, Oxford, Ohio, USA in June 1976.

In 1981, he took up a course in 'Marine Resources' at the University of Hawaii at Manoa, by which a lecture on 'ocean thermal energy conversion' was given by the late Prof Dr. John P. Craven [https://en.wikipedia.org/wiki/John_P._Craven]. During the period 2006–2008, as part and parcel of the Malaysian Marine Survey of the southern part of the South China Sea (MyMRS), the temperature profiling of the water column in the Northwest Trough or rather Sabah Trough has been established (**Figure 2**) where the temperature drops to 4°C at a depth beyond 1200 m, and thus, Malaysia has indeed the potential of harnessing its ocean thermal energy resources in its deep waters, deeper than 700 m or more (**Figure 3**).

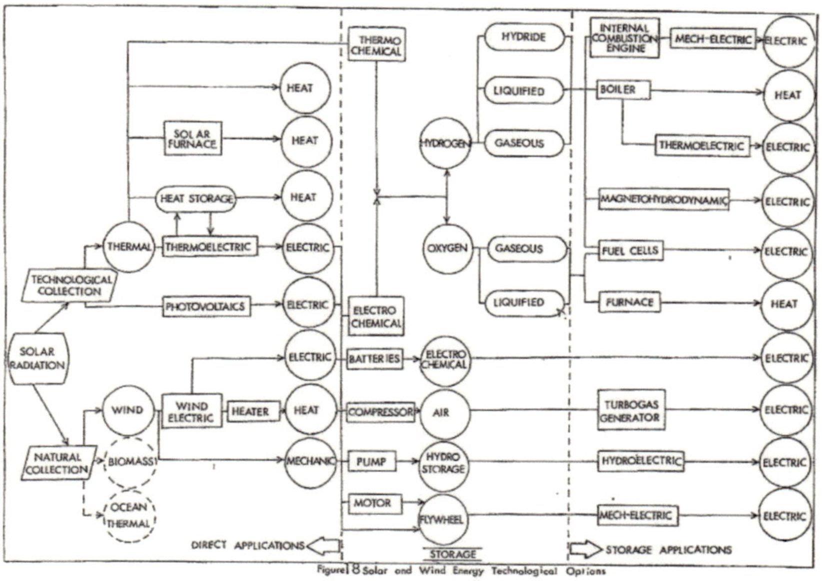

[Ref: Abu Bakar Jaafar (1976). "Applicability of Solar Energy Technology for Industrial Pollution Control and Production: The Case of the Primary Copper Smelting Industry". An Internship Report. Submitted to the Faculty of Miami University in partial fulfillment of the requirements for the degree of Master of Environmental Science Institute of Environmental Sciences. Oxford, Ohio. P.82]

Figure 1.
Solar thermal, wind, ocean thermal energy conversions to hydrogen and other energy carriers.

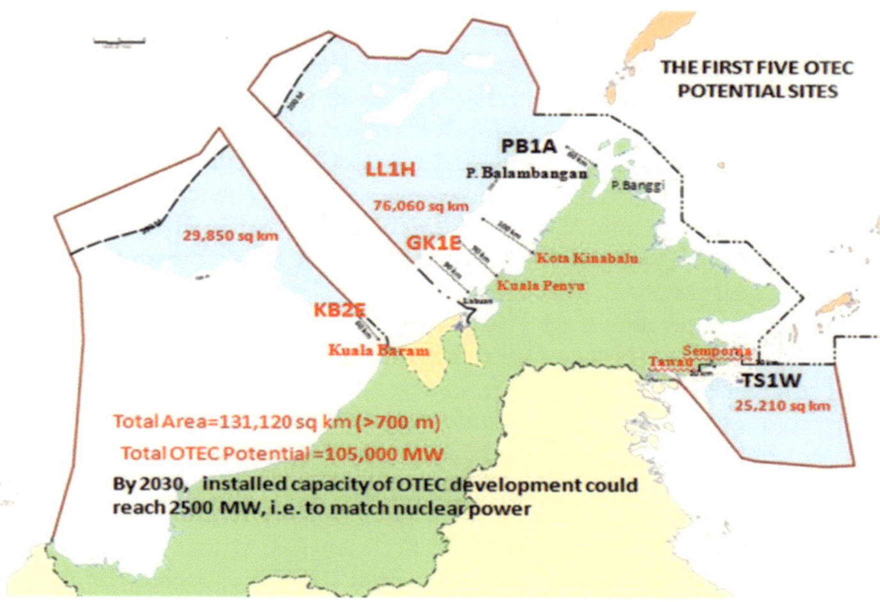

Source : Tan Ah Bah (JUPEM) & A Bakar Jaafar (MIMA & UTM Ocean Thermal Energy Centre)

Figure 2.
Malaysia: ocean thermal energy resource potential in waters deeper than 700 m. Note: the estimation of OTEC potential energy conversion is 800 kw/km², as recommended by Begamudre [1]. The potential harnessing area is evaluated based on the site of a minimum 700 m water depth. Total OTEC Potential = OTEC Potential Energy Conversion (P) x Potential Harnessing Area (A).

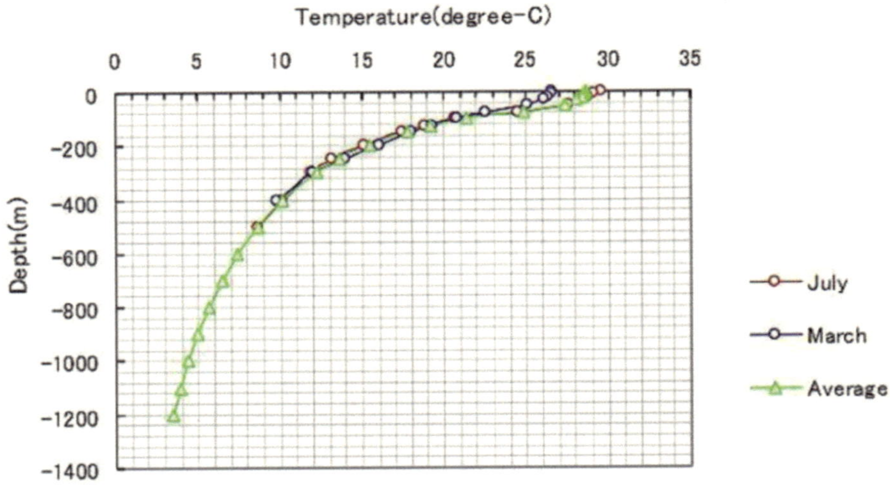

Figure 3.
Variation of seawater temperature with depth at Sabah Trough.

Since then this potential has been promoted among a number of key government agencies (including National Committee on Continental Shelf of the Secretariat of the National Security Council (MKN); Ministry of Science, Technology, and Innovation (MOSTI); Ministry of Energy, Green Technology, and Water (KeTTHA); Maritime Institute of Malaysia (MIMA); Malaysian Industry-Government Group for High Technology (MIGHT); Malaysia's Performance Management and Delivery Unit (PEMANDU); Minister for the Economic Planning Unit, Prime Minister's Department; Academy Science Malaysia-Clean Energy Panel; Malaysian Institute of Economic Research (MIER); Office of the Science Advisor, Prime Minister's Department, Malaysia; Ministry of Entrepreneur Development), potential investors (such as UTM Holdings SdnBhd; Chemicals Company of Malaysia Bhd (CCMB); PASDEC Holdings Bhd; PETRONAS; Lobsters Aquatech Sdn Bhd; FELDA Investment Corporation; Maser Marine Sdn Bhd; Lembaga Tabung Angkatan Tentera (LTAT); Arab Construction Corp (ACC); FELCRA; ANGKASA; APEX; Sabah Economic Development and Investment Authority (SEDIA); Technip Malaysia; FIDES Advisory; Affin-Hwang Capital; MCD Venture Holdings; Brunei Darussalam Economic Development Board; Chiyoda Malaysia; MOFAZ; Master Pack Bhd; Infrakomas Sdn Bhd; Sabah Rural Development Corporation; Black Pearl Energy Sdn Bhd; Prisma Simfoni Sdn Bhd), and other stakeholders (**Figure 4**) and presented at numerous fora, seminars, workshops, conferences, and symposia (i.e. Centre for the Environment, Technology, and Development, Malaysia; UPM Conference on Climate Change; UM 2nd South China Sea Conference; UUM ICOIS 2013: The first International Conference On Innovation and Sustainability; MOSTI: 1st Ocean Renewable Energy Symposium; 1st (Honolulu), 2nd (Goseong), 3rd (Kuala Lumpur), 4th (Amsterdam), 5th (La Reunion), 6th (Okinawa) International OTEC Symposia; MICCI-28th CACCI Conference; EC 6th National Energy Forum; IEM-IET Conference; EUMCCI).

On 3 January 2013, Universiti Teknologi Malaysia (UTM) established its UTM Ocean Thermal Energy Centre (UTM OTEC). The main purpose of this Centre is as much to promote and commercialise (C) the emerging ocean thermal energy conversion (OTEC) technology as to carry out further development (D) and research (R) (CDR Approach).

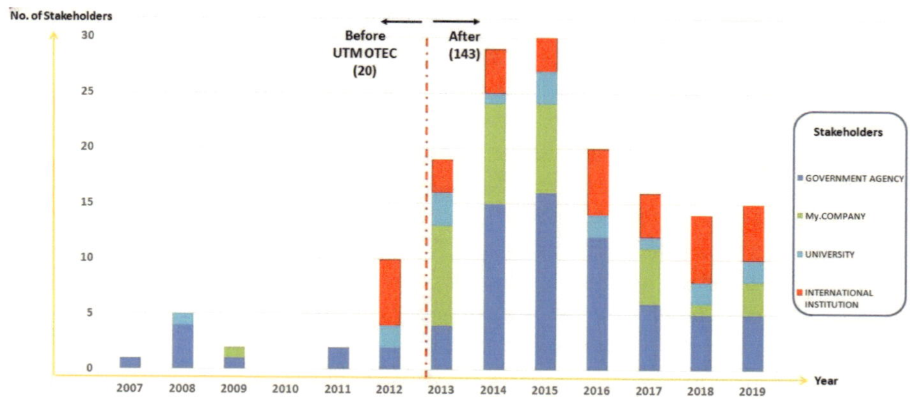

Figure 4.
Malaysia, Ocean Thermal Energy-Driven Development: growth in activities relating to OTEC Policy Advocacy and Promotions of Investment (2007–2019) as on 23 August 2019.

Prior to this, arising from the IOC-WESTPAC Workshop on Marine Renewable Energy held in Melaka, Malaysia on 17 February 2012, a close contact has been made between Prof Dr. Yasuyuki Ikegami, Director, the Institute of Ocean Energy of Saga University, Vice-Admiral (R) Shin Tani, former Chief of Hydrography of Japan, and Prof Dato' Ir Dr. A. Bakar Jaafar, Visiting Fellow of the Maritime Institute of Malaysia (MIMA).

Presentations on OTEC Malaysia at the Secretariat of the Headquarters for Ocean Policy (SHOP), Cabinet Secretariat, the Government of Japan; and to Prof Dr. Tetsuro Urabe's graduate students at the University of Tokyo, and series of Briefings on the Institute of Ocean Energy of Saga University @Imari Plant, and Okinawa Prefectural Deep Sea Water Research Center, Mayor Office of Kumejima, Bade Haus Spa, Point Pyuru Cosmetics, Kumejima Kaiyoushinsousui Kaihatu Co Ltd., Yumairu Museum, Xenesys, Inc., and New Energy and Industrial Technology Development Organisation (NEDO), and not last, Meeting with Mr. Nagamine, Director-General, and Ms. Kato, Director of Ocean Division, International Legal Affairs Bureau, Ministry of Foreign Affairs, Japan.

Upon a successful visit to Japan for various discussions and presentations on OTEC potential and future developments in Malaysia (i.e. at the Secretariat of the Headquarters for Ocean Policy (SHOP), Cabinet Secretariat, the Government of Japan; and to Prof Dr. Tetsuro Urabe's graduate students at the University of Tokyo, and series of Briefings on the Institute of Ocean Energy of Saga University @Imari Plant, and Okinawa Prefectural Deep Sea Water Research Center, Mayor Office of Kumejima, Bade Haus Spa, Point Pyuru Cosmetics, Kumejima Kaiyoushinsousui Kaihatu Co Ltd., Yumairu Museum, Xenesys, Inc., and New Energy and Industrial Technology Development Organisation (NEDO), and not last, Meeting with Mr. Nagamine, Director-General, and Ms. Kato, Director of Ocean Division, International Legal Affairs Bureau, Ministry of Foreign Affairs, Japan), upon the official invitation of the Secretariat of the Headquarters for Ocean Policy (SHOP), Cabinet Secretariat, the Government of Japan, over the period of 1–7 July 2012, Prof Dato' Ir Dr. A. Bakar Jaafar, after having joined the Universiti Teknologi Malaysia, as a Visiting Professor at the UTM Perdana School of Science, Technology, Innovation, and Policy, convened on 10 July 2012 a Workshop on Ocean Thermal Energy Conversion (OTEC) attended by over 15 UTM Academic and Administrative Staff. The Workshop put up a strong recommendation that UTM ought to establish its programme relating to OTEC.

Prior to this, on 24 April 2012, in Mansassas, Virginia, USA, Prof Dato' Ir Dr. A. Bakar Jaafar accompanied Mr. Arham Abdul Rahman, Director of MIDA, New York met Mr. Jeremy Feakinsof Chairman of The Board at Ocean Thermal Energy Corporation, and his senior Officials. The former managed to have made an appointment for Mr. Feakins and his technical advisor, Dr. Ted Johnson to meet up with YAB Prime Minister of Malaysia, Dato' Seri Mohd Najib Tun Abdul Razak, on 16 May 2012 at Mandarin Oriental Hotel, Mid-Manhattan New York. The Corporation was interested to invest in an OTEC Project in Malaysia but wanted to know what would be the amount of royalty payment to be imposed and the relevant laws applicable. As an outcome, a Cabinet paper was drafted by MIMA after having successfully conducted an Inter-Agency Stakeholders Consultation Workshop on 10 May 2012, and presented it to an ad-hoc Inter Agency Steering Committee Meeting on 20 June 2012, and to the Minister of Transport since then.

3.2 Post-2013 activities

While continuing its promotion for investment in OTEC, UTM Ocean Thermal Energy Centre managed to secure a number of research and study grants, including the following:

i. UTM-PDRU of MYR 65,000.00 [23 February 2015–2022 February 2016]: "Techno-Economic Analyses of OTEC-RELATED Industries" [2].

ii. UTM-GUP of MYR 49,650.00 [1 May 2015–2026 January 2017]: "Solar Thermal Chilled Water Systems for Temperate Agriculture" [3].

iii. UTM-FRGS of MYR 71,000.00 [2 November 2015–1 December 2017]: "The Regulatory Framework on the Implementation of Ocean Thermal Energy-Driven Development within Malaysian Waters" [4].

iv. UTM-OTEC & DCNS-Naval Energies Pre-Feasibility Studies of OTEC Project off Pulau Layang-Layang under the Ministry of Defence, Defence Industry and Ministry of Finance, Technology Development Agency (TDA) Offset Programme 2016–2017 [5];

v. UTM-SF of MYR 274,300.00 [1 May 2015–2031 July 2017]: "Integration of Rankine Power and Absorption Refrigeration Cycles for Low Load Optimized Solar Thermal Chilled Water Soil Cooling System" [6].

vi. UTM-PDRU of MYR 75,350.00 [1 July 2017–2030 June 2018]: "DNA Barcoding for Marine's Flora and Fauna Keystone Species in Pulau Layang-Layang Sabah" [7].

vii. UTM-JASTIP of JPY300,000.00 [16 October 2017–2030 September 2018]: "Conceptual Design of Fixed Ocean Thermal Energy Conversion Offshore Power Plant". This is a joint research collaboration between Universiti Teknologi Malaysia and Saga University (Japan) funded by the Japan-ASEAN Science, Technology and Innovation Platform (JASTIP-Net 2017). The outcome has been presented at the 5th JASTIP Symposium, Kuala Lumpur, Malaysia.

viii. UTM-Infrakomas Sdn Bhd Joint Pre-Feasibility Studies of OTEC Project off PulauWeh, Aceh, Indonesia of MYR 322,195.80 [1 April 2017–2031 March 2018 extended to 31 December 2019]. This is a University-Industry collaboration program which provided the financial support to Mohammad Shafiq Rahmat, contract Research Officer, who continued his PhD programme on eDNA of Deep Waters under the supervision of DrSuriyanti Su Nyun Pau, former Post-Doctoral Fellow, Senior Lecturer of Universiti Kebangsaan Malaysia (UKM).

ix. UTM-GUP of MYR 50,000.00 [1 February 2018–2031 January 2020]: "Conceptual Design of Fixed Ocean Thermal Energy Conversion (OTEC) Offshore Power Plant in Malaysia". This project is a University-Industry collaboration research programme, namely, between Universiti Teknologi Malaysia, Sarawak SHELL Berhad, and Technip FMC Malaysia Sdn Bhd. The primary outcome from this programme is an operable design of the first fixed offshore OTEC pilot power plant consist of single-legged structure and dual-level production decks. This initiative allows OTEC technology becomes more viable by massive reduction of the capital expenditure (CAPEX) of OTEC facility compared to the current OTEC Floating plant.

x. UTM-LRGS&SATREPS-OTEC Project of MYR 6,000,000 under the Ministry of Education of Malaysia + USD4,000,000 in kind from Japan Science and Technology Agency (JST) for the period 1 July 2019–2030 June 2024: "Development of Advanced Ocean Thermal Energy Conversion Technology for Low Carbon Society and Sustainable Energy System: First Experimental OTEC Plant in Malaysia".

3.3 Annual workshops

The UTM Ocean Thermal Energy Centre has been organising its annual or special workshops in 2016 [8], 2017 [9], and 2018 (**Figure 5**) as part and parcel of its objectives as much to promote the subject of 'ocean thermal energy' as to explore further opportunities for research and investment. A special workshop [10] on ocean energy was held in December 2017, jointly with the EU-Malaysia Chamber of Commerce and Industry.

3.4 3rd International OTEC Symposium (IOSKL2015)

The UTM Ocean Thermal Energy Centre convened the 3rd International OTEC Symposium [11] at UTM Kuala Lumpur Campus on 1–2 September 2015.The

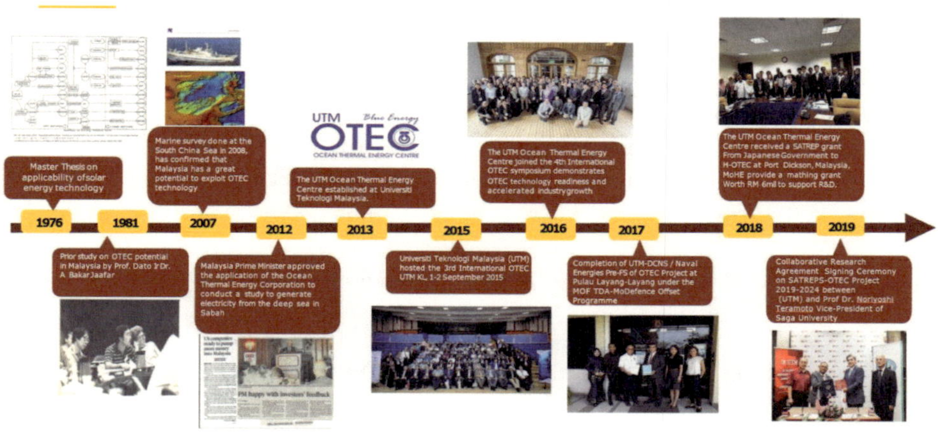

Figure 5.
UTM Ocean Thermal Energy Centre: key activities prior to and post 2013.

3rd IOSKL2015 saw 200 OTEC experts, speakers, and participants from all over the world including France, Indonesia, Japan, Korea, Netherlands, Philippines, Singapore, Sudan, Sweden, USA and Malaysia. There were 15 topical presentations at plenary, two videos, and 20 paper presentations during two parallel workshop sessions. All contributions have provided an up-to-date knowledge sharing among the researchers and set the direction to the OTEC community for future research and development including commercialization.

3.5 International exchange of young researchers

The Institute of Ocean Energy (IOES), Saga University, Japan has been sponsoring two young researchers below 35 of age from UTM OTEC to participate at the Annual Program of International Platform on Ocean Energy for Young Researcher since 2015. The researchers of UTM OTEC who had participated were as follows: Nor Amyra Hana Mohd Yusoff and Mohd Alshafiq Tambi Chik (1st Program 2015), Nur Hidayah Nong Nazari and Samihah Zura Mohd Nani (2nd Program 2016), Mohd Khairi Abu Husain and Nurul Syazwani Mohd Sabri (3rd Program 2017), Suriyanti Su Nyun Pau and Nurul Azizah Mukhlas (4th Program 2017), and Mohammad Shafiq Rahmat and Meng Soon Chiong (5th Program 2018). Among them, Nurul Syazwani has won the Best Presentation Award at the 3rd Program of International Platform on Ocean Energy for Young Researcher in 2017.

3.6 JST-MoE SATREPS project (2019–2024)

On 2 July 2019, at Dewan Seminar UTM KL, Universiti Teknologi Malaysia (UTM) and Saga University of Japan entered into the Collaborative Research Agreement (CRA) on the Development of Advanced Ocean Thermal Energy Conversion Technology for Low Carbon Society and Sustainable Energy System: First Experimental OTEC Plant in Malaysia. This Agreement would essentially cover the provisions of an experimental rig that allows for the running and testing of stainless-steel evaporator with warm sea-surface water that has gone through flash distillation chamber in order to eliminate totally the problems of biofouling of the said heat-exchanger; introduction of various designs of 3 kW turbo-generator that would run on ammonia and other alternative working fluids, including nanofluids (ORC system with different combination of nanoparticle infused liquid (nano-liquid) as the working medium), plant operational performance analysis, techno-economic analysis, the generation of chilled-water system in improving the culture of high value fish and other marine products, including seaweeds, the like of umi-budou. The quality of surface and deep waters would also be characterised, including the determination of its contents of flora and fauna by eDNA technique.

Essentially, there would be a set of 10 sub-projects, under the Development of Advanced Hybrid Ocean Thermal Energy Conversion (OTEC) Technology for Low Carbon Society and Sustainable Energy System: First Experimental OTEC Plant of Malaysia, as follows:

i. Facility: [led by Prof Dato' Ir Dr. A-Bakar Jaafar, FASc of Universiti Teknologi Malaysia];

ii. Operational: [led by Ts. Dr. Sathiabama T. Thirugnana of Universiti Teknologi Malaysia];

iii. Heat Exchanger: [led by Dr. Chiong Meng Soon of Universiti Teknologi Malaysia];

iv. eDNA: [led by Dr. Suriyanti Su Nyun Pau of Universiti Kebangsaan Malaysia];

v. Turbine: [led by Prof Ir Dr. Shuhaimi Mansor of Universiti Teknologi Malaysia];

vi. Seaweed: [led by Prof Dr. Phang Siew Moi, FASc of University of Malaya];

vii. Fish: [led by Prof Dr. Aziz Arshad of Universiti Putra Malaysia];

viii. Sea Water Quality: [led by Prof Dr. Fatimah Md. Yusoff, FASc of Universiti Putra Malaysia];

ix. Nanofluids: [led by Assoc. Prof Dr. Nor Aswadi Che Sidik of Universiti Teknologi Malaysia]; and

x. Business Development: [led by Assoc Prof Dr. Mas Bambang Baroto of Universiti Teknologi Malaysia].

The experimental rig of the said hybrid-OTEC System and related-research would be installed and conducted at the Universiti Putra Malaysia (UPM)-International Institute of Aquaculture and Aquatic Sciences (I-AQUAS), Port Dickson, Negeri Sembilan, Malaysia.

4. Commercialization of ocean thermal energy conversion technology

Prior to and beyond the establishment of UTM Ocean Thermal Energy Centre in 2013, it has been quite a challenge to attract an anchor partner to undertake the development of the first Pioneer Ocean Thermal Energy-Driven Development Project in Malaysia. The first five [12] promising potential sites have been identified and promoted to various parties or stakeholders, including deep-water production of oil & gas operators the like of PETRONAS; FELDA Investment Corporation Sdn Bhd; FELCRA Holdings Sdn Bhd; MoD Defence Industry; and Sabah Economic Development & Investment Authority (SEDIA).

The pre-feasibility study of the first site at Pulau Layang-Layang, conducted jointly between UTM OTEC & DCNS Naval Energies under the MoD Defence Industry-MoF TDA Offset Program was completed on 12 September 2016, and jointly accepted on 27 September 2019 at the 4th International OTEC Symposium, Amsterdam, the Netherlands.

In-house pre-feasibility studies also completed for two other sites: (i) 15 km off Pulau Kalumpang, Tawau-Semporna, Sabah, Malaysia, and (ii) off Pulau Banggi, Kudat, Sabah. It is quite likely the first site would be taken up by a local engineering group at the end of 2019.

A regional investor, based in Singapore, has set up a local company, and is seriously looking into the prospects of supplying OTEC power to the existing and future deep-water production not only in Malaysia, but also in Brunei Darussalam and Indonesia.

The fifth deep water site off Kuala Baram had been offered to the State Government of Sarawak, but the subject has been deferred simply because Sarawak has big surplus of hydro-power. This surplus power would be converted to hydrogen fuel, as the State of Sarawak has embarked toward 'hydrogen economy' by having installed the first hydrogen fuelling station in Kuching on 27 May 2019, and by deploying the first three hydrogen-fuelled buses, and two standard utility vehicles [13].

5. Future prospects

The best strategy for realising the potential of OTEC Technology is as much to convert the excess power into hydrogen fuel as to produce drinkable mineral water from the deep sea off the OTEC Plant condensers. There are other revenue streams that could be developed, the like of Kumejima OTEC Park in Okinawa, Japan, and NELHA Kona OTEC Park in Hawaii, USA, in the production of high value seaweeds, fugu fish, prawns, abalone, and temperate cash-crops. There is no other renewable energy conversion technology that could match OTEC, not even salinity gradient, being the lowest cost of power producer with the least investment per unit power produced, as highlighted in **Figure 6** and **Table 1** of an article [14]:

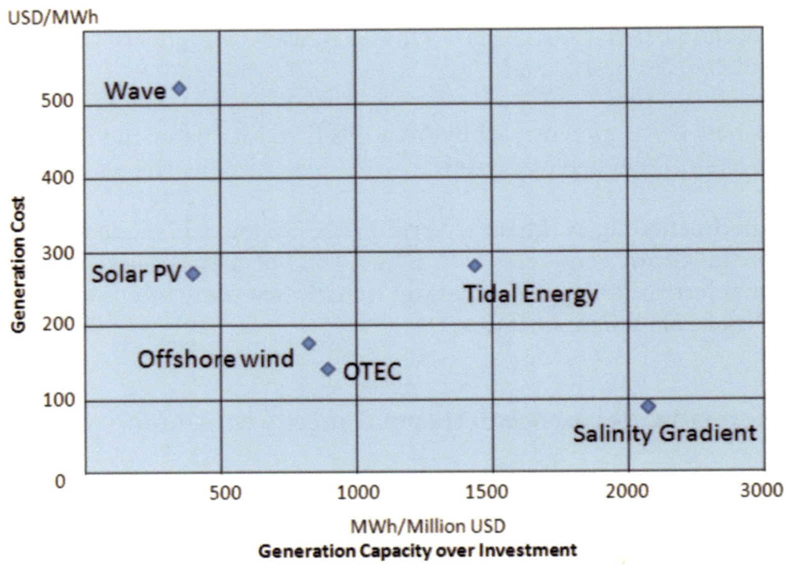

Figure 6.
Comparative analysis of various forms of renewable energy (RE) [14].

Ocean Energy and Solar PV	Input					Capacity factor	Output	
	Generation Capacity (MW)	Annual Power Generation (MWh)year	Capital Investment (Million USD)	Generation Capacity/Capital Investment (MW/Million USD)	Annual Power Generation/Capital Investment (MWh/Million USD)		Cost of Ocean Energy (USD/KWh)	Cost of Ocean Energy (USD/MWh)
Wave Energy	10	24,000	63	0.16	380	30%	0.56	560
Tidal Energy	254	406,400	298	0.85	1363	20%	0.28	280
Offshore wind	10	33,600	40	0.25	840	42%	0.17	170
OTEC	53	402,800	451	0.12	893	95%	0.13	130
Salinity gradient	200	1,280,000	600	0.33	2133	80%	0.09	90
Solar PV	10	16,000	38	0.26	421	20%	0.25	250

kWh: Kilowatt-hour, MW: Megawatt

[Data Source: IRENA (2012, 2014), ADB Report (2015)]

Note: **Table 1** *compares three economic variables that impact feasibility: the capital investment of energy, which is the start-up cost for a project, the net capacity factor, which is the ratio of actual energy output in comparison to full potential of energy generated and the levelized cost of ocean energy (COE), which is the cost to build and operate an energy generation device over its lifetime divided by the total power generated over its lifetime. The ideal energy resource has a low COE and capital cost and a high capacity factor [15].*

Table 1.
Cost of renewable energy generation by capital investment and capacity [14].

Although the amount of capital required for investment in OTEC is high, as shown in **Figure 6**, the generation cost of electricity is the lowest, other than that of salinity gradient, compared to those generated from wave, wind, oceanic current, biomass, tidal current, and solar photovoltaic [16–22]. It is simply because, as outlined in **Table 1**, the capacity factor of OTEC is at least 95%, compared to that of tidal current, only 20%, and where its capital requirement is the lowest per unit of power generated.

OTEC power would become more competitive, than that is stated above, when its total capital investment is spread out across its spinoffs, and that could be covered from the sales of raw deep seawater for mineral water production, marine culture of high value produce, growing of temperate crops. Such OTEC spinoff industries have been well developed.

Acknowledgements

This is to thank all former fellows, associates, students and staff who had been part and parcel of UTM Ocean Thermal Energy Centre for their invaluable contributions. This chapter would not have been completed without the continuing support of the current staff, students and fellows. This is also to thank the management of Universiti Teknologi Malaysia, faculty members and staff for their support too.

Author details

A. Bakar Jaafar*, Mohd Khairi Abu Husain and Azrin Ariffin
UTM Ocean Thermal Energy Centre, Universiti Teknologi Malaysia, Kuala Lumpur, Malaysia

*Address all correspondence to: bakar.jaafar@gmail.com

References

[1] Begamudre RD. Energy Conversion Systems. New Delhi: New Age International (P) Ltd.; 2000

[2] Banerjee S, Musa MN, Jaafar AB, Ariffin A. Application on solar, wind and hydrogen energy–A feasibility review for an optimised hybrid renewable energy system. Journal of Fundamentals of Renewable Energy and Applications. 2015;5:193. DOI: 10.4172/20904541.1000193

[3] Zain NEM. Experimental simulated soil conditions for temperate crop growing in lowland areas in Malaysia [Master thesis]. Malaysia: Malaysia-Japan International Institute of Technology (MJIIT);

[4] Rani MHA. The regulatory framework on the implementation of ocean thermal energy-driven development within Malaysian waters [PhD thesis]. In: UTM OTEC Proceedings 3rd National Workshop on Ocean Energy 2017. Malaysia: Malaysia-Japan International Institute of Technology (MJIIT); Available from: https://otec.utm.my/prooceedings/

[5] Zahar M. Marine microbial diversity of off-terengganu coastal sediment in South China sea [PhD thesis]. Malaysia: Malaysia-Japan International Institute of Technology (MJIIT); 2017

[6] Rasaq Adekunle O, Jaafar AB, Musa MN, Saad AA. Adapting temperate crops to tropical lowland through solar thermal chilled water soil cooling process. Applied Mechanics and Materials. 2016;**818**:231-236. Available from: https://doi.org/10.4028/www.scientific.net/amm.818.231

[7] Sabri NSA. Role of temperature in soils for the growth of temperate crops [PhD thesis]. Malaysia: Malaysia-Japan International Institute of Technology (MJIIT); 2019

[8] UTM-OTEC Proceedings 2nd National Workshop on Ocean Energy. 2016. Available from: https://otec.utm.my/blog/2016/10/12/proceedings-2nd-national-workshop-on-ocean-energy-2016/

[9] UTM-OTEC Proceedings 3rd National Workshop on Ocean Energy. 2017. Available from: https://otec.utm.my/prooceedings/

[10] UTM OTEC-EU Proceedings EUMCCI Seminar on Ocean Energy

[11] UTM OTEC Proceedings IOSKL. 2015. Available from: https://otec.utm.my/prooceedings/

[12] Bakar Jaafar A et al. Framework for OTEC development in Malaysia. In: 1st International OTEC Symposium, Honolulu Convention Center, Oahu September 9-10. 2013

[13] Available from: https://www.theedgemarkets.com/article/sarawak-launches-aseans-first-integrated-hydrogen-production-plant [Accessed: 22 August 2019]

[14] Bakar Jaafar A. Renewable ocean thermal energy-driven development for sustainability. The Ingenieur. 2019;77:26-30

[15] Navigant, prepared for: U.S. Department of Energy. 2014. Offshore Wind Market and Economic Analysis

[16] International Renewable Energy Agency. 2014. Ocean Thermal Energy Conversion Technology Brief. Available from: http://www.irena.org/DocumentDownloads/Publications/Ocean_Thermal_Energy_V4_web.pdf

[17] International Renewable Energy Agency. 2014. Renewable Energy Technologies: Cost Analysis Series: Wind Power. Available from: http://

www.irena.org/DocumentDownloads/Publications/RE_Technologies_Cost_Analysis-WIND_POWER.pdf

[18] International Renewable Energy Agency. 2014. Renewable Power Generation Costs in 2014. Available from: http://www.irena.org/DocumentDownloads/Publications/IRENA_RE_Power_Costs_2014_report.pdf

[19] International Renewable Energy Agency. 2014. Salinity Energy Technology Brief. Available from: http://www.irena.org/DocumentDownloads/Publications/Salinity_Energy_v4_WEB.pdf

[20] International Renewable Energy Agency. 2014. Tidal Energy Technology Brief. Available from: http://www.irena.org/DocumentDownloads/Publications/Tidal_Energy_V4_WEB.pdf

[21] International Renewable Energy Agency. 2014. Wave Energy Technology Brief. Available from: http://www.irena.org/DocumentDownloads/Publications/Wave-Energy_V4_web.pdf

[22] International Renewable Energy Agency. REMap 2030- Renewable Energy Roadmap. United States of America: Renewable Energy Prospects; 2015. Available at: http://www.irena.org/remap/IRENA_REmap_USA_report_2015.pdf

Chapter 7

General Criteria for Optimal Site Selection for the Installation of Ocean Thermal Energy Conversion (OTEC) Plants in the Mexican Pacific

Alejandro García Huante, Yandy Rodríguez Cueto, Erika Paola Garduño Ruiz and Ricardo Efraín Hernández Contreras

Abstract

The purpose of this chapter is to provide an assessment of the resource potential for ocean thermal energy conversion (OTEC) in the Mexican Pacific Ocean (MPO). Research methodology adopted in this study is a combination of geographic information system (GIS), to identify the most promising site in the MPO for OTEC deployment. Site selection criteria rely on conditions such as distance to cold water pumping, bathymetry, thermal difference (not less than 20°C), and social and environmental aspects. Finally we concluded that sites located in the states of Guerrero and Oaxaca have the highest potential of the entire MPO, although there are other areas in the states of Baja California Sur, Nayarit, or Michoacan that might have some interest for OTEC technology.

Keywords: OTEC, sites, Mexican Pacific Ocean, thermal difference

1. Introduction

Mexico has public policies that allow the implementation of renewable energy (RE) projects. One of them is the energy reform that indicates the market incentives of clean energy certificates which will be defined by 2019.

Therefore, it will be obligatory for all the companies that generate energy to have obtained or bought these clean energy certificates and the settled goals in the Energy Transition Law (LTE, from its name in Spanish) [1]. This standard establishes a minimum share of clean energy in electricity generation of 25% by 2018, 30% by 2021, and 35% by 2024 [1].

The necessity of developing the RE sector in Mexico has led to government institutions, as the Ministry of Energy (SENER, from its name in Spanish) and the National Council for Science and Technology (CONACyT, from its name in Spanish), to create Mexican Centers Energy Innovation (CeMIEs, from its name in

Spanish). CeMIEs are groups of public or private research, higher education institutions, companies, and government entities that have the objective of working together on projects dedicated to developing technologies, products, and services, which allow Mexico to take advantage of its enormous potential in the main RE. For development of technologies for ocean energies, the CeMIE-Océano (CeMIE-O) was recently set up which focuses on the assessment of wave, current, tides, saline gradient, and thermal gradient energy resources in Mexico [2].

The Mexican Pacific Ocean (MPO) and the Caribbean Sea (CS) are areas which are an excellent thermal potential for OTEC, according to several studies [3–8], which can reduce fossil fuel dependency and provide energy to isolated coastal populations and islands.

The main criteria to determinate optimal site or zones are their location to be less than 10–15 km offshore from the Mexican coastline and the thermal difference (TD) to be equal to or greater than 20°C. Furthermore, other criteria are the distance to the cold water pump area that should not exceed 10 km (for onshore plants) as well as the social and economic factors that exist in these communities.

Finally, the importance of this technology is to provide electricity in the isolated coastal areas as well as other subproducts generated, such as fresh water, seawater air-conditioning (SWAC), cold agriculture (ColdAg), and aquaculture, among others. That is why it is also imperative to calculate the net electric generation given by a particular area and the contribution to social development, the interest of the Mexican and foreign energy sectors in this type of power generators, and the scientific, social, economic, oceanographic, and climatic contribution that can be generated [9].

2. Methods and data

2.1 Study area

The MPO is the study area, as the geographic and oceanic conditions are suitable for the installation of OTEC plant. In this area, the TD between 0 and 1000 m depths and the seasonal variations of the sea surface temperature (SST) were analyzed by [9, 10]. The authors found 43 sites, which meet the afore mentioned main criteria. Sites are located in the Gulf of California (GC), Baja California Sur (BCS); Eastern Pacific Warm Pool in the Pacific Ocean (ETP), Nayarit (NYT), Jalisco (JAL), Michoacán (MCN), Guerrero (GRO), and Colima (COL); and Gulf of Tehuantepec (GT), Oaxaca (OCX).

2.1.1 Temperature

Strong change of temperature in shallow waters called thermocline characterized Eastern Tropical Pacific (ETP) [11–13] (see **Figure 1**). ETP is influenced by "El Niño" (ENSO), which is important in climate variability, in fish production, and in the global carbon cycle [14]. The annual variation in the SST fluctuates between 26 and 30°C [3, 14] with variations of flow by some cold currents as the California Current from the north and the Peru Current from the south. The intrusion of the Costa Rica Coastal Current (CRCC) causes annual variations of 5°C or more in the area near JAL, while, in the GT, the temperature varies between 3 and 4°C (see **Figure 1**).

Databases used for this work to obtain the data include of the Mexican Navy (SEMAR, from its name in Spanish), the National Oceanographic Data Center (NODC) of the United States of America, and the World Ocean Database (WOD)

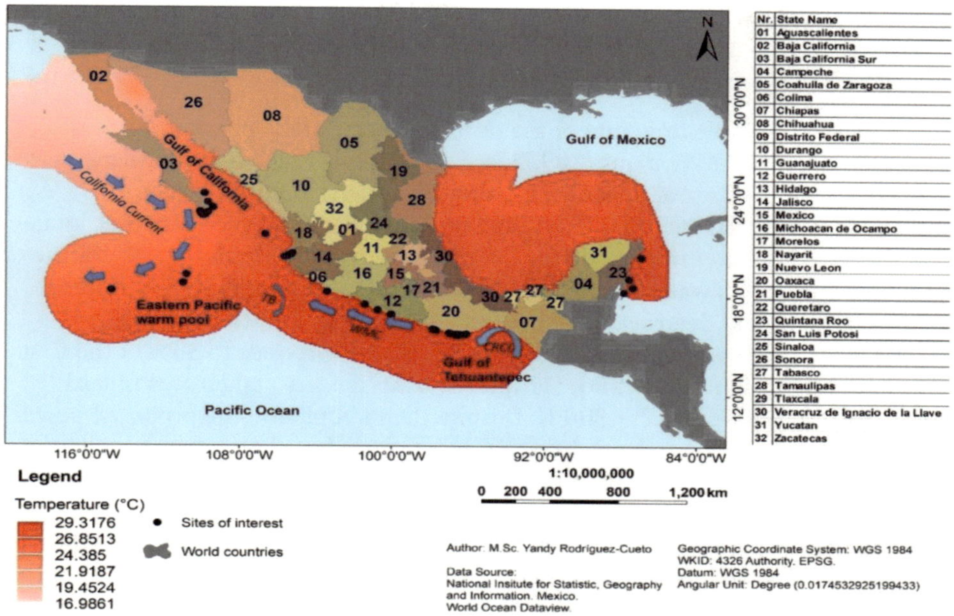

Figure 1.
Eastern tropical Pacific (ETP) and ocean circulation patterns. Shading represents mean sea surface temperature (darker = warmer). Nomenclature: TB (Tehuantepec bowl), WMC (Western Mexican current), and CRCC (Costa Rica coastal current).

2013, of the National Oceanographic and Atmospheric Administration (NOAA), through the Ocean Data View (ODV). Historical and seasonal data of temperatures in the MPO was compiled in order to determine the changes of oceanic TD occurring over time. From the corresponding vertical profiles of temperature, the thermal gradient was computed and plotted.

2.1.2 Analysis and interpretation of data by geographic information systems (GIS)

GIS tools allowed obtaining datasets like TD, which is the result of subtracting SST from the temperature reported for other depths. The exclusion process allows for representing those areas with TD above 20°C. The MPO was divided into three regions: the GT, the ETP, and the GC. Subsequently, the sites with suitable thermal gradients for the operation of an OTEC plant were determined, which will then be plotted on different maps.

3. Net electric power

3.1 Methodology

Nihous [15] proposes a method to estimate the efficiency conversion of thermal energy to power generation for a device (100 MW theoretical plant). In this method, he and his coworkers assumed that the TD is distributed in main parts of this power plant in a structure called "temperature ladder." The formula proposed is based on a single thermal energy generation device and is expressed as:

$$P_{net} = Q_{CW} \frac{3\rho C_p \varepsilon_{tg} \gamma (\Delta T)^2}{16(1+\gamma)T} \tag{1}$$

$$P_{pump} = Q_{CW}\, 0.30\, \frac{\rho C_p \varepsilon_{tg} \gamma}{4(1+\gamma)} \tag{2}$$

where $\boldsymbol{Q_{cw}}$**:** Volume flow rate of the deep seawater intake per unit (latitude-longitude) for a 100 MW plant (309.7174 $m^3\ s^{-1}$); $\boldsymbol{\rho}$**:** Average seawater density as 1025 kg m^{-3}; $\boldsymbol{C_p}$**:** seawater-specific enthalpy equivalent to 0.004 MJ $kg^{-1}\ K^{-1}$; $\boldsymbol{\varepsilon_{tg}}$**:** turbo-generator efficiency as 0.75; $\boldsymbol{\gamma}$**:** flow rate ratio of the surface seawater intake over the deep seawater intake (1.5 for floating systems); **T:** absolute temperature of the OTEC surface seawater intake (K = °C + 273.15); **ΔT:** temperature difference between surface and deep seawater intakes (°C).

The pumping power density defined in Eq. (2) corresponds to 30% of the first term in the right-hand side of Eq. (1) (gross power density) taken at standard conditions ΔT = 20°C and T = 300 K. Despite their simplicity, the proposed power formulas capture the basic behavior of OTEC systems based on pure-fluid, low-temperature Rankine cycles, with thermodynamic efficiencies approximately half the Carnot limit of ΔT over T [15].

With the calculations obtained, net power maps were prepared for the MPO area, both annual and seasonal. With this, sites of interest were determined for the application of this technology and compared with other power values for other types of energy (thermoelectric, photovoltaic, wind, etc.), in order to know the scope of the OTEC cycle and what percentage of the national market could it have at a given time if it can be applied.

4. Social and environmental factors

4.1 Methodology

To describe the importance and impact of social, economic, and environmental factors of sites with potential for OTEC, you must know the variables that allow us to know what the marine environment is like for life, historical information from the PO. DAAC [16] was used to have a general notion that was subsequently compared with information from Marine Copernicus [17] and CONABIO [18] and adjust subsequent effort and investment to characterize and have direct measurements at the sites of greatest interest to OTEC. The marine variables related to water chemistry that may be related to their interaction with living beings.

The availability of water is of the utmost importance for life and the economic development of any region of the world. Information was considered in a superficial principle, but later it was suggested that knowing what depths of which it is intended to extract cold water could give greater knowledge of the oceanic conditions existing in the Mexican seas for these variables or parameters. Salinity, chlorophyll (mg m^{-3}), nitrates (μmol L^{-1}), phosphates (μmol L^{-1}), dissolved oxygen (mL L^{-1}), and silicates (μmol L^{-1}) were used, and the variables were possible at 0, 500, 700, 750, and 1000 m depth.

Finally, social research was done in three regions of Mexico: Baja California Sur (La Paz), Nayarit (Punta Mita), and Oaxaca (Puerto Ángel) where surface temperature measurements have also been made considering the sites of interest for implementation of this technology. The importance of this field study is to know the ideology of the people about the problems existing in their communities and their knowledge about marine renewable energy.

5. Results

5.1 Mean annual historical thermal difference (TG) maps (0–1000 m depth)

Throughout the year at 1000 m depths, it is observed that the TD increases considerably from Nayarit to Oaxaca where optimal TD is around from 22 to 24.5°C in the ETP and GT. The TD at GC, which includes sites as Baja California Sur, varies from 19 to 22°C although seasonal data showed the most representative TD value was below 20°C. From the previous information, the MPO is a favorable area for the installation of OTEC (see **Figure 2**).

5.2 Net electric power

Based on the analysis performed, the net electric power maps (annual average and seasonal) were obtained. The average annual electrical power map shows that the southern and southeastern parts of the Mexican Pacific tend to have a greater amount of energy for most of the year regardless of the type of OTEC cycle used. Guerrero and Oaxaca would produce almost 200 MW of electricity in a year, which translates into about 1500 GWh year^{-1} (see **Figure 3**).

The southern part of the peninsula of Baja California, the Gulf of California, and part of the central Mexican Pacific Ocean would generate powers of between 100 and 140 MW per year. As the OTEC cycle does not require the burning of any fuel for evaporation of the working fluid, the electricity cost is considerably reduced, and, with the secondary products that can be obtained from the cycle (drinking water, air conditioning, aquaculture water, and production through cold water agriculture, among others), populations could enjoy benefits of labor and environmental and social development, which would result in a decrease in rates such as violence and poverty in the country. Additionally, these sites could have their

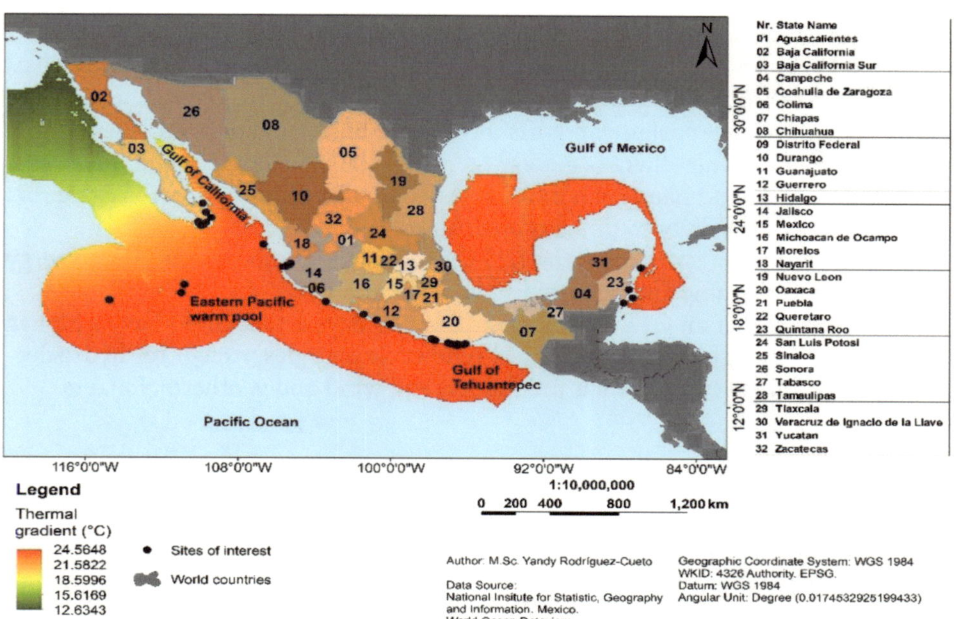

Figure 2.
Mean annual historical (1955–2012) thermal difference between 0 and 1000 m depth.

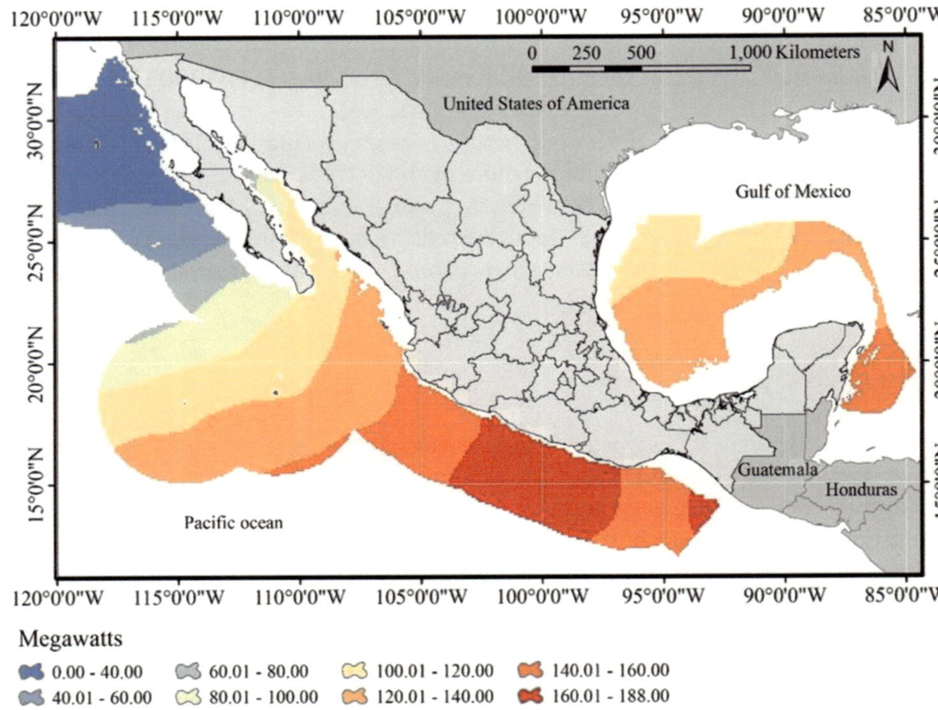

Figure 3.
Mean annual historical (1955–2012) net electric power in the Mexican Pacific Ocean.

energy needs covered so that federal investments in this sector would fall considerably and thus the money could be used in other important social sectors.

However, the OTEC cycle is very sensitive to changes that may exist in the sea, so variations in temperature differentials will generate distinct electric power production. As can be seen in **Figure 3**, there are substantial changes in net power according to seasonality, reaching values ranging between 60 and 200 MW, the best sites for the operation of the plant being the states of Guerrero, Oaxaca, and Chiapas, where plants onshore and offshore can be built and operated (these last plants will operate when there is greater technological development and improvement of structures and parts such heat exchangers, pipes, and submarine cable for the transport of electricity to the port area, among others).

In **Table 1**, the comparison of the OTEC production numbers with other energy sources, both renewable and nonrenewable, is observed.

It is observed that an OTEC plant operating 365 days a year could have a higher average net energy generation than most renewable energies, except for hydroelectric plants that have twice the net production and wind and geothermal plants.

5.3 Environmental factors

If the environmental conditions that are present in the sites that can be considered optimal for establishing an OTEC are well known, it will help in the choice of materials, construction and maintenance of the equipment, and infrastructure and thus have a better planning and idea of costs. Considering installing pipes and pumping deep water to the surface of the sea and then reinjecting at an intermediate depth will modify the marine and coastal environment to some extent. Until the moment the plant is operating it will be possible to evaluate and have a clear

Energy type	Total (MW)	Plant factor	Average anual production net (GWh year^{-1})	Net hours of annual production
OTEC[2]	**100**	**0.8**	**1562.5**	**8760**
Hydroelectric	2400	0.41	3498.1	3555
Solar/Phootovoltaic	30	0.46	19.5	1415
Wind	251.1	0.32	225	2801
Geothermal	225	0.77	1172.3	6767
Bioenergy	53	0.24	26.7	2100
Nuclear power	1608	0.75	7925.8	6572
Carboelectric	2778	0.73	12897.6	6360
Thermoelectric	2100	0.37	2488.7	3203
Dual	367	0.64	1145.5	4877
Turbogas	393	0.28	274.4	2494
Internal combustion	210	0.25	113.5	2162
Regenerative brakes	290	0.56	1071.1	6596
Combined cycle	1454	0.67	5650.2	5800

[1]*Source: SENER (prepared with data from the CFE, CRE, CENACE, SEMARNAT, and the Secretariat of Planning and Energy Transition, 2016). In blue, renewable energy. In red, conventional energies.*
[2]*Theoretical.*

Table 1.
Effective installed capacity (MW) of renewable and conventional energy plants[1] compared to the theoretical OTEC cycle, average annual net production, plant factor, and net production hours.

estimation of costs considering the impact that could have and knowing the ocean dynamics that occurs can help mitigate or minimize consequences.

5.3.1 Temperature

This parameter is the most important for the operation of an OTEC plant because the duty cycle works from the Carnot cycle, which implies that there should be a preferential thermal differential of 20°C or greater, although in some countries such as South Korea, it works with a differential of 18°C in the summer months.

In the case of temperature, by analysis that we have performed at the level of numerical modeling and by results generated in some of the OTEC experimental plants, it is known that this parameter would not generate a thermal pollution process because the discharge water would stabilize once it is placed in the depth whose density is adequate.

Thus, as there is no risk of a possible thermal effect, there is no strong impact on the organisms in the water column. However, it is considered that there should be frequent monitoring of this and other parameters to be completely sure.

An example of this is the analysis performed at one of the sites of interest in Mexico for the implementation of OTEC (Puerto Angel, Oaxaca) by Instituto de Ciencias del Mar y Limnología (ICMyL) of Universidad Nacional Autónoma de México (UNAM) [4], where, according to the data obtained and after analysis (taking as a reference the summer and winter months), it is considered that the operation of the OTEC plant may not generate thermal contamination because the vertical temperature profiles, both without operation and with the operation of the OTEC plant, are similar; the above implies that the discharge water (whose output temperature is $15.5 \pm 1.7°C$ when mixed) will tend to be located in the water column, according to its density.

5.3.2 Salinity

Salinity is an environmental factor of great importance because it is a factor that defines the organisms that can adapt to living in a body of water and resist between certain ranges of salinity. The concentrations of K^+ and Cl^- are relatively conservative, that is, they undergo small variations induced by environmental changes or by their use by living beings. In contrast, the concentrations of HCO_3^{2-}, SO_4^{2-}, and Ca^{2+} depend on precipitation-dissolution processes, microbial metabolism, and climatic changes, including the sea surface temperature. Some minor elements, such as silicon, nitrogen, phosphorus (P), and iron, are of great importance from a biological point of view, although they do not significantly influence the salinity of a body of water.

In the previous example (Puerto Ángel, Oaxaca), this parameter does not have much relevance during the operation of an OTEC plant, but it is considered since it has implication in the density of water bodies. What is observed in both cases (both without operation and with the operation and operation of the OTEC plant) is that the variation in salinity is minimal (between 0.1 and 0.2 UPS) and its tendency is almost linear from 200 m deep.

This implies that the discharge water of the plant would generate small changes in salinity, its density being similar to the waters surrounding the discharge zone, so its effect would be limited.

5.3.3 Chlorophyll (primary productivity)

Chlorophyll is an indicator of primary productivity that occurs in the seas. Chlorophyll a is the compound with the greatest presence in photosynthetic organisms; therefore, the chlorophyll is directly related to plant biomass, to determine the deterioration or balance of an ecosystem.

5.3.4 Nutriments

As a result of the analyses carried out through the CORMIX program by García-Huante [4] at the Institute of Marine Sciences and Limnology regarding the behavior of nutrients in the discharge water of a 100 MW OTEC plant in Puerto Ángel, Oaxaca, it was obtained that all the nutrients would have a high concentration at 113 m depth, which is where the water would end up being located due to its density. Next are the results per nutriment.

5.3.5 Nitrates

Nitrites and nitrates are part of the nitrogen cycle. Nitrate is used as a nutrient by photosynthetic organisms, and its presence in water can accelerate the phytoplankton overgrowth, producing water eutrophication. Nitrite levels higher than 0.75 ppm in water can cause stress in fish, and greater than 5 ppm can be toxic. Nitrate levels between 0 and 40 ppm are generally safe, and any value greater than 80 ppm can be toxic.

In this case, during CORMIX simulation of the operation of an OTEC plant in Puerto Ángel, two possible increases in this nutrient are observed both in summer and in winter. By summer, nitrates would increase to 43.42 μmol L^{-1} to 113 m deep; in winter it would increase to 40.3 μmol L^{-1} to 113 m deep, which could imply an important effect.

5.3.6 Phosphates

In the sea, phosphorus is found as a phosphate $(PO_4)^{2-}$ ion that is an indispensable element in the synthesis of organic matter in the sea and is widely used by phytoplankton; its shortage in photosynthetic activity zones limits primary productivity and all marine life. Its concentration in the sea is very variable seasonally because it depends on fluctuations in the phytoplankton population.

In this case, during CORMIX simulation of the operation of an OTEC plant in Puerto Ángel, there would also be a possible increase in the concentrations in a depth of 113 m from 0.3 to 3.7 μmol L^{-1} during the summer; meanwhile in winter there would also be an increase from 0.3 to 3.8 μmol L^{-1} to 113 m deep, so the effect could be high.

5.3.7 Silicates

In seawater, the silicates are used by diatoms in the form of opal silica to form their frustules, which are like cases that keep them protected from the outside. Diatoms as organisms provide 20% of photosynthesis. Sponges, silicoflagellates, and radiolaries use it as a structural component. In animals, it is part of blood and bones, and in plants, and it protects against biotic stress and abiotic stress.

During CORMIX simulation of operation of the OTEC plant in Puerto Ángel, there are also two possible sites where the concentration of silicates would increase in both summer and winter. For the first case, the increase would be 91.24 μmol L^{-1} 113 m deep; in winter it would be 101.7 μmol L^{-1} at 113 m deep. The effect, from this increase, would be significant.

However, the dilution rate of these nutriments and the surface currents that exist in the area would end by elimination in a matter of 2–3 weeks, according to the simulation performed. Therefore, there would be no high risk in addition to the fact that the discharge water would be located below the photic layer (which in the area usually reaches 100 m depth) depending on the seasonality and the oceanographic processes that occur in the area as the upwelling or ENSO.

Hence, according to the analysis carried out by García-Huante [4], the impact of oceanographic dynamics on the nutrients of the discharge water of the OTEC plant is decisive; it is unlikely that there will be more activity at the trophic chain level. Thus, some type of involvement, both positive and negative, could be ruled out at the biological level in the study area.

5.3.8 Dissolved oxygen

In the sea, oxygen is a necessary element for all life forms. Natural stream purification processes require adequate oxygen levels to facilitate aerobic life forms. The dissolution of oxygen occurs between the surface of the water and the atmosphere, and many factors interfere. It is also produced as a residue of photosynthesis, but it is quickly consumed and depleted in the degradation of organic compounds.

During CORMIX simulation of the operation of OTEC plant in Puerto Ángel, the oxygen would not have negative changes because the poor waters of the bottom, at the time of being pumped, would be oxygenated by the turbulent mixing action both in the extraction and by the passage in the pipes of plant. Based on previous studies (Guenther and Green, 1990, and Vega, 1995), this parameter would not have a negative effect on the environment.

5.4 Social factors

The term OTEC is an unknown technology in the national scene and the experience that, as a working group, have had with the coastal communities in explaining how it works and what it implies. The response observed in the population is generally optimistic, especially in the communities with the greatest lag and the desire that its implementation can bring social benefits that they seek.

The social and economic perspectives of Mexico are different in any region suggested to build an OTEC plant. However, it will be wrong to think that the social benefit will be to provide electricity at no cost, jobs, or money as income for the communities.

In Puerto Ángel, Oaxaca, a community dedicated to artisanal fishing, the aforementioned thought is the prevailing one, followed by the idea that new people will arrive to mobilize their tourism and commerce. They are discouraged that the plant would sell its electricity produced to the CFE or that a private sector will manage it. The community of Puerto Ángel have the fear that strangers who arrive will break their structure and stability by bringing new problems or aggravating those the community have, and they feel that what comes from the government is often negative or against them, but on the other hand, they look at OTEC plant as a window of opportunity to reduce social and economic lag for people who are more optimistic. We talk to the community about the technological applications that can result from implementing an OTEC plant, such as obtaining cold water for cooling, drinking water if desalinated, and salts for chemical industry or agriculture.

In Bahía de Banderas, which the states of Jalisco and Nayarit share, the diversification of economic activities and services that exist in the community causes very different perspectives, ranging from indifference to enthusiasm, about what OTEC represents.

On the other hand, in Baja California Sur, fisherman community and people who live on the coast have a different perspective, while their economy depends on the sea without necessarily being fishermen, such as a merchant, shipyard, and mechanic, supplies for fishing, or maintenance of boats or cars. When that community listen to about the description of OTEC, they ask technical things, for example, what type of pumps will be used, their resistance to corrosion or inclement weather, the profit of generating potable water and electricity using this technology, and application of cold water in refrigeration, agriculture, and aquaculture production.

We are concerned to solve the environmental impacts of reinjection of seawater and how the OTEC plant could modify the energy matrix of the community before deciding on a particular site.

6. Discussion

6.1 Thermal difference

In order to know optimal sites of OTEC installation in MPO, it is necessary to evaluate each one with physical, environmental, socioeconomic, and legal aspects with a biggest vision that allows us to know the advantages and disadvantages of future OTEC practices.

The regions that seem to be promising and optimal for OTEC are ETP and GT. Both areas present the TD favorable throughout the year. There are various sites of great depth close to the coast, having possible access to the electrical grid, and there are important centers of consumption. Among the unfavorable aspects, there is the possible threat of intense hurricanes and the fact that there are protected areas.

Concerning hurricanes, on this side of the Pacific Ocean, they evolve from south to north in such a way that, while in the south the hurricane category is normally 1 or 2, category 5 hurricanes strike the north and even travel as far as the Baja California peninsula.

Two sites are currently under intensive study: Bahía de Banderas, Nayarit, and Puerto Angel, Oaxaca. In the case of the Mexican Pacific Islands (MPI), the areas with the maximum TD are observed near the Marias and Marieta Islands, which are protected areas. In addition, the Islas Marias contains a prison, so that any ocean engineering projects would require special permission from the Mexican government.

6.2 Power net energy

The best places to operate an OTEC plant should be those where the temperature is quite stable. However, the plant can adapt to existing oceanographic conditions so that regions of the central Pacific and southern Mexico can use OTEC without major risk, unlike the southern part of Baja California where values fall considerably and would have an operating time between 4 and 6 months.

This analysis indicates that, under an annual average production of 200 MW of electricity (3125 GWh $year^{-1}$) and with a plant factor of 0.8 (80% operation, 365 days a year and 24 h a day), an OTEC plant could have a higher average net energy generation than most renewable energies and very close to hydroelectric production. As for conventional energy, OTEC is competitive in some cases in terms of production, but not enough because there are some other energy sources that present a much higher production as the hydroelectric, nuclear, and fuel plants (see **Table 1**).

However, the secondary products obtained from OTEC plants give it a higher rating when making a possible decision.

6.3 Environmental and social factors

First, regarding the possible environmental factors, it is important to consider water pollution, due to chemicals emitted by machine parts, and to evaluate the geological characteristics of the sites in case of erosion or earthquake. Then, the electromagnetic effects of power lines and the possibility of marine life attaching onto the devices are worth examining. Finally, it is crucial to verify the effects on protected zones and to perform predictions regarding the occurrence of hazards that are common in Mexico, such as storms and hurricanes.

Last, in the case of social factors, it is important to promote the approval of the community for the activities to be carried out, preserving the culture and traditions of the region. It is also very important to consider areas, which are potential zones of recreation, transportation, aquaculture, or military activities. Moreover, it is crucial to count on academic, government, and industrial participation in the projects. Finally, security factors should be considered where the sites are affected by human migration, criminal activities, and so on.

7. Conclusions

MPO has an excellent thermal potential for OTEC deployment. The most promising regions into MPO are the ETP and the GT, which includes sites from Nayarit to Oaxaca and the MPI. Sites located in the Gulf of California have inadequate gradient for part of the year; nevertheless, other applications such as the utilization of cold water can be of great interest. The communities in these areas would have a high social development expectation, considering the amount of secondary by-products derived from the cycle, the decrease in the cost of electricity production (and therefore in the payment for the service), as well as achieving greater stability to eliminate gradually violence and poverty in these places.

The social perception that broadly reflects the surveys made to the inhabitants of the area of interest for the CEMIE-O denotes some discomfort; infrastructure development linked to strengthening sectors of the tourism industry has been characterized, according to fishermen's point of view, for only taking into account economic and technical criteria. The repercussions that have had decisions such as the relocation of an entire locality, waste management, and some environmental protection interventions are issues that the social groups in question try to solve without many tools at their disposal.

Although there are cases in which the omissions to the social dimension have not escalated into major conflicts, an issue can completely stop large-scale projects. The comprehensive account of the social dimension is that one must have careful planning and a genuine interest in the structure and ways of life of the community in question. This is not only a requirement to be met in the administrative order of project management but the basis of sustainable development as a concept and as a practice.

Acknowledgements

The authors would like to thank the National Council for Science and Technology (CONACyT), the Mexican Center of Innovation in Ocean Energy (CEMIE-O), and the Institute of Marine Sciences and Limnology (IMCyL) of the National Autonomous University of Mexico (UNAM) for supporting this research.

Conflict of interest

The authors declare no conflict of interest. The founding sponsors had no role in the design of the study; in the collection, analyses, or interpretation of data; in the writing of the manuscript; and in the decision to publish the results.

Appendices and nomenclature

SENER	Ministry of Energy
CONACyT	National Science and Technology Council
CEMIE-O	Mexican Energy Innovation Center of Ocean
RE	renewable energy
LTE	Energy Transition Law
OTEC	Ocean Thermal Energy Conversion
MPO	Mexican Pacific Ocean

CS	Caribbean Sea
GIS	Geographic Information System
EEZ	Mexican Exclusive Economic Zone
TD	thermal difference
SST	sea surface temperature
GC	Gulf of California
BCS	Baja California Sur
ETP	Easter Pacific Warm Pool in the Pacific Ocean
NYT	Nayarit
JAL	Jalisco
MCN	Michoacán
GRO	Guerrero
COL	Colima
GT	Gulf of Tehuantepec
OCX	Oaxaca
El Niño	ENSO
CRCC	Costa Rica Coastal Current
SEMAR	Mexican Navy
NODC	National Oceanographic Data Center
WOD	World Ocean Database
NOAA	National Oceanographic and Atmospheric Administration
ODV	Ocean Data View
MPI	Mexican Pacific Islands
CFE	Electricity Federal Commission
ICMyL	Institute of Marine Sciences and Limnology
UNAM	National Autonomous University of Mexico
CRE	Energy regulatory commission
CENACE	National Energy Control Center
SEMARNAT	Ministry of Environment and Natural Resources

Author details

Alejandro García Huante[1*], Yandy Rodríguez Cueto[1], Erika Paola Garduño Ruiz[1] and Ricardo Efraín Hernández Contreras[2]

1 Instituto de Ingeniería, Centro Mexicano en Innovación en Energía Oceánica (CeMIE-O), Universidad Nacional Autónoma de México, Mexico City, Mexico

2 Instituto de Ciencias del Mar y Limnología, Centro Mexicano en Innovación en Energía Oceánica (CeMIE-O), Universidad Nacional Autónoma de México, Ciudad Universitaria, Mexico City, Mexico

*Address all correspondence to: alex_dodo@hotmail.com

References

[1] Ministry of Energy (SENER) [Internet]. May 2018. Available from: https://www.gob.mx/sener/articulos/el-fondo-para-la-transicion-energetica-y-el-aprovechamiento-sustentable-de-la-energia-es-un-instrumento-de-politica-publica-de-la-secretaria

[2] CEMIE-Océano. [Internet]. 2019. Available from: https://cemieoceano.mx/

[3] Avery WH, Wu C. Renewable Energy from the Ocean. A guide to OTEC; Oxford university press: Oxford, UK; Research Triangle Institute: Durham, NC. In: USA. 1994. DOI: 978-1-60119-684-2

[4] García Huante A. Posibles Efectos Oceanográficos por la Operación de una Planta OTEC en Puerto Ángel,Oaxaca, México [Master's Thesis]. Posgrado en Ciencias del Mary Limnología. Mexico City: Universidad Nacional Autónoma de México; 2015

[5] Crews R. OTEC Sites, Aquarius Rising Maldives—An Ocean Research Centre and Eco-Tourist Facility. 1997

[6] Aviña H. Factibilidad de Instalar una Planta OTEC (Ocean Thermal Energy Conversion) en las Costas de México. Bachelor's Thesis: Universidad Nacional Autónoma de Mexico, Mexico City, Mexico; 2007

[7] de la Cerda V, Boletín Informativo ESIA IPN May-June R. IPN-ESIA. Mexico: Mexico City; 1984

[8] Francisco BGJ. Evaluación del Potencial de Conversión de Energía Renovable en el Caribe Mexicano [Master's Thesis]. Mexico City: Universidad Nacional Autónoma de México; 2014

[9] Alejandro GH, Yandy RC, Rodolfo S, Edgar M, Luis VA. Determination of the potential thermal gradient for the Mexican Pacific Ocean. Journal of Marine Science and Engineering. 2018; **6**:1-11. DOI: 10.3390/jmse6010020

[10] Cerdeira Estrada S, García Huante A, Garduño Ruiz EP, Martell Dubois R, Rodríguez Cueto Y, Valdéz J, et al. System of Information and Analysis of the Potential for Energy Use of the Ocean on the Coasts of Mexico Based on OTEC-Mex (OceanER) IINGEN-CONABIO. CONABIO. México: Information & Analysis System of Marine Ecosystem of Mexico (SIMAR); 2019. https://simar.conabio.gob.mx

[11] De la Lanza-Espino, G. Gran Escenario de la Zona Costera y Oceánica de México. Ciencias. 2004;**76**:4-13

[12] Huante-González, Y. Contaminación Biológica en la Bahía de Puerto Ángel, Oaxaca. 1997. Available from: http://www.umar.mx/revistas/2/contaminacion.pdf ()

[13] Alatorre Mendieta MÁ, Hernández Contreras RE, García Huante A, y Rodríguez Cueto Y. Atlas de Potencial Energético del Océano en México por Gradiente Térmico; Línea Estratégica de Energía por Gradiente Térmico del Centro Mexicano en Innovación de Energías del Océano; CEMIE-Océano: Mexico City, Mexico, 2017

[14] Kessler WS. The circulation of the eastern tropical Pacific: A review. Progress in Oceanography. 2006;**69**: 181-217

[15] Rajagopalan K, Nihous GC. Estimates of Global Ocean thermal energy conversion (OTEC) resources using an ocean general circulation model. Renewable Energy. 2013;**50**: 532-540

[16] PO.DAAC Physical Oceanography Distributed Active Archive Center.

2019. Available from: https://podaac-tools.jpl.nasa.gov/las/UI.vm

[17] Marine Copernicus Open Access Hub. 2019. Available from: http://marine.copernicus.eu/services-portfolio/access-to-products/

[18] CONABIO Geoportal. 2017. Available from: http://www.conabio.gob.mx/informacion/gis/

Chapter 8

Environmental Impact Assessment of the Operation of an Open Cycle OTEC 1MWe Power Plant in the Cozumel Island, Mexico

Enrique Celestino Carrera Chan,
María Fernanda Sabido Tun, Juan Francisco Bárcenas Graniel and Estela Cerezo Acevedo

Abstract

An environmental impact assessment (EIA) was made for the operation of a 1MWe open-cycle OTEC plant on Cozumel Island, Quintana Roo. Due to its bathymetric characteristics and its proximity to the population center, the ideal location for the placement of the OTEC plant is in the western coastal area of the island. An environmental inventory was developed in which the susceptible factors to be impacted were described (air, soil, water, landscape, geology and flora). The essential components of the OC-OTEC PLANT operation process were studied: vacuum pump, flash evaporator, turbine, condenser and pipes. An impact matrix (Leopold matrix) was created, which prompts to impacts on the environment list, generated by the OC-OTEC PLANT operation: CO2 emissions; nutrients dragged to the surface; artificial reef effect/contamination by heavy metal salts; drag and compression of organisms; redistribution of oceanic water bodies; impacts by organic antifouling chemicals; noise; illumination; turtle nesting alteration; brine discharges; cause significant public controversy; alteration and interruption of migration routes; and waste, among others. Once the most significant impacts were assessed through the Leopold matrix, corrective and preventive measures were established on those actions, in order to minimize their negative impact on the environment.

Keywords: environmental impact assessment, open-cycle OTEC, Leopold matrix, Cozumel

1. Introduction

Energy is a conserved physical quantity, and the first law of thermodynamics states that energy can be transformed from one form to another but can be neither created nor destroyed. It is under this precept that we can argue that in nature there are endless ways to obtain energy by transforming it from a primary source.

A common source of sustainable energy is the ocean, which has six categories of energy generation: waves, tides, tidal currents, ocean currents, ocean thermal conversion and salinity gradient [1]. Taking into account the large proportion of water that exists on the planet, oceans can be considered as a large energy storage system.

Ocean Thermal Energy Conversion (OTEC) uses the difference in temperature (thermal gradient) between the surface of the sea (the hot spot) and the waters of the deep sea. Mexico has a high thermal energy gradient; therefore, it is in a favorable location for ocean energy use, through the implementation of OTEC plants [2].

Currently, studies exist and research has been conducted on the useful energy of OTEC plants, both for open cycle (OC) and closed cycle (CC). These lack a deep development in the estimates of environmental impact of construction, running and/or decommissioning of projects of this magnitude. Therefore, it has not been possible to determine, predict or interpret the impacts that could result as an effect from the anthropogenic activities related to the operation of an OTEC plant.

This project aimed to demonstrate that the operation of an open-cycle 1MWe OTEC plant generates relevant environmental impacts (positives or negatives) on environmental factors such as air, soil, geology, water, landscape, flora, population and fauna, corresponding to a polygon of the coastal zone western of Cozumel Island.

This project is part of an agreement between Universidad del Caribe and Centro Mexicano de Innovación en Energía del Océano (CEMIE-Océano) belonging to the research line called "Detección de lugares de aprovechamiento de gradientes de temperatura con potencial energético de explotación en México para determinar la factibilidad de construcción de una planta tipo OTEC (GLE-1)" (Detection of places of use of temperature gradients with energy potential of exploitation in Mexico to determine the feasibility of building an OTEC type plant (GLE-1)).

2. Methodology

The methodology used for the environmental impact assessment (EIA) of the operation of the OC-OTEC plant project is an adaptation of the one proposed by the author Gómez Orea (1998) [3], which has 4 phases in which the environmental impacts resulting from the interaction of activities that are developed theoretically during the extraction of the operating water resource of an OC-OTEC PLANT are identified, characterized, assessed and interpreted, as well as the rejection of water that comes from the plant and that goes toward ocean mass and other secondary activities that are broken down during maintenance and operation. This will be evaluated in a polygon of 1096 m^2 located in the western coastal area of Cozumel Island, Quintana Roo, the four phases of this methodology are mentioned below.

2.1 Phase 1: description of the project

During this phase, the bases covered by the scope of the general vision of the project were raised and fleshed out. A zone was defined where the OC-OTEC plant would be theoretically located. Chosen based on a series of parameters that would suit the operation of the plant, two alternative sites for the project were registered that due to state environmental legislation had to be discarded, as well as an ethnographic study that aided in the knowledge of public acceptance, which allowed us to measure the reception the project would have based on the perception of the inhabitants of Cozumel Island. The points described above, as well as others, are recorded in **Table 1**.

2.2 Phase 2: description of the project environment and environmental inventory

The description of the project's surroundings allows the categorization of the area that will be affected directly and indirectly by the activities of the OC-OTEC

No	Parameters to be described
1	Name of the project
2	Head of the project/responsible person or agency
3	Location
4	Production process and manufactured products
5	Other possible locations
7	Activities that might have impact
8	Jobs created in the operation phase
9	Public acceptance

Table 1.
Parameters for phase 1.

plant in such a way that each environmental factor was delimited by a series of geographic borders in order to narrow the research space. In this way, polygons were drawn according to the different areas of involvement.

Similarly, an environmental inventory was prepared so that those factors that could be transformed during the operation phase of the open-cycle OTEC plant will be described. For this, an analysis of the natural environment will be made and the possible effects will be contrasted once the OTEC plant enters into operation; in this way, the magnitude of the impact will be known.

Table 2 lists factors that will be described during the environmental inventory.

Medium	Environmental factor	Parameters for inventory
Inert	Air	Winds
		Hurricanes
		Weather
		Acoustic contamination
	Soil	Type of soil
		Potential land use
	Geology	Geology
	Water	Ocean water
		Surface water
		Ground water
		Geohydrology
Perceptual	Landscape	Visual
	Flora	Flora
	Population	Demographics
		Economic activities
	Fauna	Sea fauna
		Birds

Table 2.
Factors considered for the environmental inventory.

2.3 Phase 3: identification and assessment of impacts

During this phase, an environmental impact assessment tool was used, which helped us to quantify the environmental impacts from the convergence between the actions presented during the operation stage of an OC-OTEC plant (mentioned in phase 1) with the environmental factors that were identified in phase 2. This tool is called Leopold matrix and belongs to a group called *cause-effect matrices.*

2.4 Phase 4: proposal of corrective and preventive measures

Of the actions identified as having the highest impact (described in phase 3), corrective and preventive measures will be proposed that minimize their negative impact.

3. Results

3.1 Phase 1: description of the project

The project was titled "Environmental Impact Assessment of the operation of a 1MWe open-cycle ocean thermal energy utilization (OTEC) plant in Cozumel, Quintana Roo, Mexico." CEMIE-O (Spanish acronym for Centro Mexicano de Innovación en Energía del Océano), the promoter of this project, is a multidisciplinary entity that includes a large number of researchers, companies and government institution. CEMIE-O focuses on the generation of knowledge and its application in different research lines, in order to raise and offer information about various forms of use of energy from the ocean.

As a reference area for the project, a polygon of 1096 m^2 corresponding to that of a 1MWe OTEC plant located in India [4] was established, which required to be a vegetation-free space that could be leveled and compacted according to the standards and construction specifications. This site was located in the western coastal area of Cozumel Island, located at km 3, South Coast, South Hotel Zone, of San Miguel de Cozumel, belonging to the municipality of Cozumel, Quintana Roo, Mexico.

In order to select the ideal physical positioning site of the OC-OTEC PLANT, a series of criteria were taken into account, which will be detailed below:

3.1.1 Thermal gradient

An ideal area within the Mexican Caribbean region to place an OTEC plant is Cozumel Island, Quintana Roo, as it has sufficient energy resources to take advantage of them [5].

The frame of reference that was used to know the temperatures that exist in the oceanic region of the western area of Cozumel Island was the historical database of the Naval Oceanographic Office (NAVO), who make use of temperature profiled buoys and salinity known as ARGO, which reach depths close to 2000 m [6].

The ARGO buoy with close geographical match to the area of study has the key number 4901644, with a starting trajectory that goes from the Caribbean Sea in an area near Honduras and Jamaica and reaches the coasts of Isla Mujeres, passing through the western section of Cozumel Island.

The measurement points closest to the study area are # 46 and # 47, and these measurements were made on October 21 and 25, 2014 (currently the buoy is inactive, so reference data were taken on the mentioned dates).

According to recorded temperature profiles, the surface temperatures of the measuring points 46 and 47 reach as high as 29°C, while those found at a depth of 400 m at both points oscillate around 10.5°C.

The theoretical energy conversion efficiency of an OTEC plant is 8% [7]; however, due to the entropy generated at various parts of the cycle, an efficiency of 3–4% is obtained. Taking into account that it is a relatively low percentage compared to conventional power plants, it should be noted that the energy obtainable from an OTEC plant comes from a renewable resource.

The energy conversion of the proposed OC-OTEC plant that will be found in the western coastal area of Cozumel Island will be around 6.12%, as can be confirmed with the Carnot efficiency formula (η) (Eq. 1):

$$\begin{aligned} &\text{Carnot efficiency: } \eta = \frac{T_1 - T_2}{T_1} \\ &\eta = \frac{(302.15\ \text{K}) - (283.65\ \text{K})}{(302.15\ \text{K})} \\ &\eta = 0.0612 \\ &\text{In percentage: } \eta = 0.0612{*}100 = 6.12\% \end{aligned} \tag{1}$$

where T_1 = surface temperature = 29°C = 302.15 K; T_2 = temperature at depth = 10.5°C = 283.65 K.

3.1.2 Bathymetry

The bathymetric profile corresponding to the ocean surface close to the western coast of Cozumel Island is 400 m, which is based on the vector data set in the International Bathymetric Chart of the Caribbean Sea and Gulf of Mexico, sheets 1-7 [8]. In the mentioned bathymetric line, a thermal gradient that falls within the energy efficiency range of the OC-OTEC plant can be obtained.

The distance that exists from the western coast, where the OC-OTEC plant would be located to the bathymetric line of interest, is 3.83 km; therefore, it meets the guidelines for the operation of a thermal gradient plant, which require a distance of less than 10 km.

3.1.3 Protected natural area (PNA)

The protected areas on Cozumel Island are composed of the three fundamental ecosystems that are predominant on the island environment: 1) the jungle ecosystem, 2) the coastal mangrove-dune and 3) the marine-reef zone. The area that was selected for the positioning of the OC-OTEC plant is outside the Flora and Fauna Protection Area called "La porción norte y la franja costera oriental, terrestres y marinas de la Isla de Cozumel" (The northern portion and the eastern coastal strip, terrestrial and marine of the Cozumel Island), as well as the protected natural area "Arrecifes de Cozumel" (Cozumel Reefs) [9].

3.1.4 Closeness to the population center

The polygon where the OC-OTEC PLANT will be located is 27.77 m away from the main avenue named *Rafael E. Melgar*. This avenue is the direct communication route that links to the population center.

The distance between the proposed surface of the OC-OTEC plant and the population center of San Miguel de Cozumel, Quintana Roo, is 2.92 km, and this satisfies the guidelines [10, 11] that state that the maximum distance at which an OTEC plant should be located in relation to the population center must be less than 10 km, because exceeding this distance would incur a significant loss of electrical energy in the course of transmission to the population center.

3.1.5 Urban infrastructure

3.1.5.1 Electrical energy

The supply of energy from Cozumel is provided by a subaquatic aqueduct from a substation in Playa del Carmen, which provides two radial circuits of 34.5 kV. In turn, this energy comes from the thermo-electric plant in Valladolid, Yucatan [12].

The supply comes from the continental side of Cozumel Island, which has two plants: one belonging to the substation of "Comisión Federal de Electricidad" (CFE, acronym in Spanish for Federal Electricity Commission) and the other has a private owner.

The private plant is owned by "Energía y Agua Pura de Cozumel S. de R.L. de C.V. Chankanaab," which uses a mixture of diesel and fuel oil for the generation of energy, while the substation of the CFE called "C.T.G. Chankanaab" has generators using diesel and operates during exceptional circumstances [13].

The annual electricity consumption in the municipality of Cozumel was 238.51 gigawatt-hours in 2012, which corresponds to 2.32% of consumption with respect to the peninsular zone and 0.086% with respect to the whole country. The expense for this same concept in 2012 was $ 396,722,000 [14].

Due to Cozumel Island's touristic importance, it is required to guarantee the supply of electricity [15]. By 2024, the current transmission infrastructure to the island, even with the installation of the capacitor banks proposed by 2021, will not be sufficient to cover the maximum demand, so it becomes an area of opportunity for the implementation of renewable technologies, as per example the use of energy from the ocean.

The inhabited private homes on the island of Cozumel that may have access to power correspond to 99.64% [16].

3.1.5.2 Water availability

"ACUERDO por el que se dan a conocer los estudios técnicos de aguas nacionales subterráneas del acuífero isla de Cozumel, clave 2305, Estado de Quintana Roo" (AGREEMENT by which the technical studies of national underground waters of Cozumel Island aquifer are disclosed, code 2305, State of Quintana Roo), published in 2013, mentions that the only source of supply for Cozumel Island is groundwater. The aquifer has an average annual availability of 35.341442 million cubic meters and the total annual extraction volume amounted to 8.2 million cubic meters per year on that year, of which 62.5% is destined for urban public use (wells destined for this use are found in the central part of the island), 24.1% for services, 13.3% for industrial use, and 0.1% for agricultural, domestic and watering uses.

Nearly 95.43% of the population of Cozumel Island has access to water inside their home, while the remaining 4.57% obtains it outside the home but within the land. Corresponding to water obtained by hauling, 79.45% corresponds to the wells that support obtaining water for urban public use [17].

3.1.5.3 Telephone/mail

Cozumel Island has local and long-distance services, national and international, both by operator and automatic, provided by Teléfonos de México S.A. de C.V. (TELMEX). The company has placed several booths in the downtown area, as well as the concession for the time/cost service in order to meet the extra demand in holiday seasons.

There are services that combine conventional services with shipments and collections of money orders, fax, and parcels; there is also a cable-television system station, Internet communication systems, cell phone and state-run print media.

3.1.5.4 Means of transport

Terrestrial: transport inside the city is designed based on tourist activity and is based primarily on the taxi service. The city also has a reduced bus service for the use of local population.

Marine: given its geographical condition, Cozumel Island has developed various port services that allow the berthing of tall and cabotage vessels, as well as the traffic and movement of various types of cargo. Two concessionary companies operate the regular Cozumel-Playa del Carmen passenger service and vice versa, with daily round trips that dock at the Island's Fiscal Dock, located in the center of the town, guaranteeing their service by an adequate passenger transport fleet. These same companies give the service of offers to tourist cruises that do not have the possibility of docking at the two positions of the current tourist dock. The companies are Cruceros Marítimos del Caribe, S.A. from C.V. and overseas.

The criteria explained determine that the location of the mentioned site met the ideal characteristics that would allow an optimal performance of the OC-OTEC PLANT project. On the other hand, two alternative positioning sites were analyzed, which were discarded since they did not meet these characteristics. The alternate sites (yellow) and the final location (red) can be compared in **Figure 1**.

The productive process of the operation of the OC-OTEC PLANT project in main components (vacuum pump, flash evaporator, turbine and condenser) and pipes [10] was examined for the identification of the processes performed by the subsystems headed by the main components, each one being of high importance

Figure 1.
OC-OTEC plant located in the occidental side on the Cozumel island. Source: Google Earth.

because within these the transformation or transportation processes may become environmental impacts. Production process is summarized and schematized as a flow chart in **Figure 2**.

3.2 Phase 2: description of the project environment and environmental inventory

3.2.1 Air

3.2.1.1 Winds

The Caribbean region, in general, has the influence of trade winds, which blow from the East to the Southeast, with an oscillating force between 15 and 20 knots during most of the year. During the first hours of the day and during the night, there is a land breeze due to a reduction in the speed of the trade winds. During the day, the sea breeze causes an acceleration in speed. Winds from the west sometimes occur, after the passage of a cold front or when a tropical cyclonic alteration approaches [18].

3.2.1.2 Hurricanes

A total of six hurricanes of different categories have had a place of entry to land, as well as influence on Cozumel Island, with the highest impact being Emily and Wilma in 2005, with maximum winds of 269 and 324 km/h, respectively, followed by Rina in 2011, with 174 km/h [19].

3.2.1.3 Atmospheric contamination

Municipal data in this area, as part of Plan de Acción Climática Municipal (PACMUN, Spanish acronym to Municipal Climate Action Plan), show that greenhouse gas emissions gained importance and were shown as tons of CO_2 equivalent. Cozumel takes records of a total of 176, 463 tons of CO_2 Eq. [20].

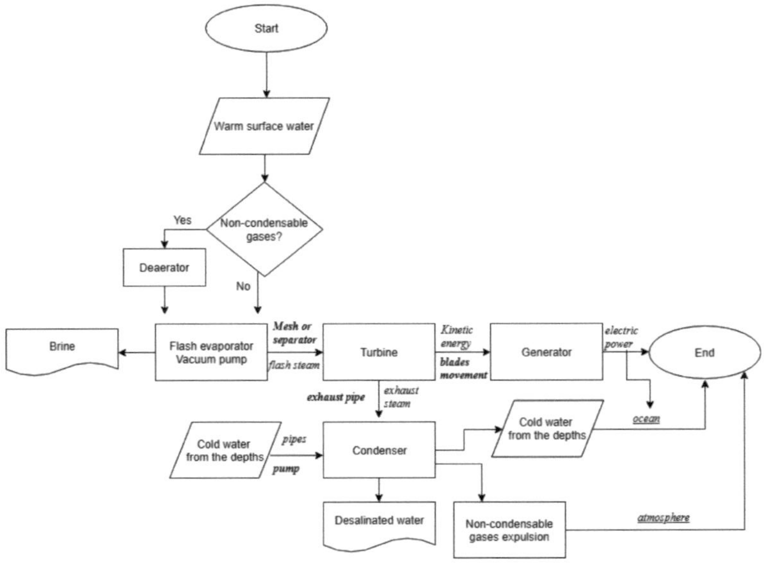

Figure 2.
Open-cycle OTEC plant process flow chart.

Of those productive sectors that contributed a relevant percentage to these emissions, the energy sector was the main contributor, followed by waste and finally agriculture.

3.2.2 Weather

The weather of Cozumel Island is warm-subhumid (Aw) with rains in summer and over 40 mm in the driest month [21]. It also has two rainfall maximums separated by two "dry" seasons, one long in the cold half of the year and one short during the rainy season [21]. The precipitation fluctuates between 27 and 45.3 mm, and the average temperature of the coldest year is 6.3° C.

The average annual ambient temperature of Cozumel island in 2016 is 29.6° C, with the month of July (during the summer season) being the hottest that year. In summer, the Caribbean Sea and the eastern Atlantic off the coast of East Africa suffer from overheating causing surface air masses to form and a decrease in atmospheric pressure, thus causing hurricane formation [22].

3.2.3 Noise pollution

Water noise measurements from a 1 MWe anchored (off-shore) OTEC-CC facility near Keahole, Hawaii, determined that seawater pumps are the maximum noise source of the OC-OTEC PLANT project, as they present a broadband spectrum that does not exceed the expected levels by more than 10 dB [23].

The corresponding Mexican official standard, "NOM-081-SEMARNAT-1994, Norma Oficial Mexicana que Establece los Límites Máximos Permisibles de Emisión de Ruido de las Fuentes Fijas y su Método de Medición", states: "*The maximum permissible noise emission level in fixed sources it is 68 dB (A) from six to twenty-two hours, and 65 dB (A) from twenty-two to six hours ...*" [24]. The operation of an OC-OTEC plant is diurnal and the value corresponding to the maximum permissible levels is not exceeded, since only 10 of 68 dB is allowed are emitted.

3.2.4 Soil

3.2.4.1 Soil type

Throughout the entire insular surface, four main groups of soils with different extensions (rendzinas, solonchak gleyico, regosol calcarico and gleysol malico) are broken down [25].

The theoretical polygon proposed for the positioning of the OC-OTEC PLANT project is on the type of rendzina soil (lithosols or leptosols as they are usually known in different classifications) that are genetically young soils and have a single horizon A thin in some. It is prostrated on a small horizon B or, if necessary, on the unchanged parental material. Leptosols have a color ranging from brown to dark red and their texture is usually clayey, and they have a basic pH (close to 8), well drained, with high water retention and lamellar stones [26].

3.2.4.2 Potential land use

This project is located on Cozumel Island, which is an Área Sujeta a Ordenamiento Ecológico (ASO, acronym in Spanish for Area Subject to Ecological Management) as stipulated by Programa de Ordenamiento Ecológico

Marino y Regional del Golfo de México y Mar Caribe (POEMRGMyMC, acronym in Spanish to Marine and Regional Ecological Management Program of the Gulf of Mexico and the Caribbean Sea), and is also located in the Unidad de Gestión Ambiental (UGA, acronym in Spanish for Environmental Management Unit) # 141, regards to the Programa de Ordenamiento Ecológico Local (POEL, acronym in Spanish for Local Ecological Management Program). This ASO prohibits the drilling of new domestic wells for water extraction from the aquifer, the disposal of wastewater in bodies of water, open-air dumps for the disposal of solid waste, burning of solid waste, the installation of any kind of fences that obstruct the movement of native wildlife, the introduction of species and the extraction of sand from beaches.

3.2.5 Geology

The Yucatan Peninsula is composed of marine compounds of calcareous composition belonging to the Tertiary and Recent period [27]. As for antiquity, the geological framework is formed by cretaceous limestone and dolomitized, solidified and recrystallized limestones belonging to the Paleocene, with clear coloration that emerge in the southern portion in the State of Yucatan and have a thickness up to several hundred meters.

3.2.6 Water

3.2.6.1 Marine waters

Cozumel Island is located in the Atlantic Ocean, in its northern part, and also in the Caribbean Sea. The semi-enclosed seas around the North Atlantic coasts have different oceanographic and bathymetric regimes, as well as ecosystems with many characteristics determined by local processes and pressures.

About waves, it is known that most of the year the winds from the East and Southeast are the dominant ones in the area, except in winter, when the direction of these changes to the North-Northwest. This causes windward coast to be the most deployed to the wave energy, causing the development of breaker sites, with steep step shape and tiny cliffs. The leeward coast is protected most of the year and is only damaged during the "nortes" season (north winds) [28].

3.2.6.2 Surface waters

Only a small fraction of the water that passes through the hydrological cycle resides in fresh surface water bodies, such as creeks and lakes [29]. Due to limestone karst erosion in the Cozumel island, there is an absence of surface water channels and an underground body of fresh water, which lies on the salty marine waters with greater density and gets maximum thickness in the central-eastern part of the island [30].

3.2.6.3 Groundwater

Comisión Nacional del Agua (National Water Commission) has divided México into 13 Regiones Hidrológicas-Administrativas (RHA, acronym in Spanish for Hydrological-Administrative Regions). Cozumel Island belongs to two RHA regions, RHA XII Yucatan Peninsula and Hydrological Region number 32 Northern Yucatan.

Wells for urban public use (which are the only source of supply for Cozumel Island) are located in the central part of the island, where the thickness of fresh water is greater. Tap water demand has increased due to the island's demographic growth as well as the growing tourism influx. Of the 486 groundwater catchments, 264 are allocated to urban public use, 59 are used in services, 21 are for industrial use, 130 for agricultural use, 2 for domestic use and 9 for the watering hole [31].

3.2.6.4 Geohydrology

The Yucatan Peninsula is mainly composed of limestones and dolomites of high permeability. That is why the high rainfall, the great infiltration capacity and the reduced topographic slope favor the renewal of the underground water of the peninsula [32].

3.2.7 Flora

The predominant flora in the polygon is the medium subdeciduous forest, which is mainly composed of two tree layers that vary between 8 and 20 m high. There is also a small shrub-herbaceous layer consisting of young specimens of the dominant species of the tree layers. The soil is poor in organic matter, as well as underdeveloped; however, there are regions of the island, especially the center, where the vegetation becomes more complex, attributed, surely, by greater land grabbing, showing a well-defined layer physiognomically and floristically, which has few climbers and epiphytes. Approximately half of the species are deciduous [28].

3.2.8 Population

3.2.8.1 Demography

According to the 2015 census, 86,415 inhabitants were reported on Cozumel Island, Quintana Roo [17]. Of the total population, men correspond to 49.27%, while women correspond to 50.73%.

Regarding the population's geographical distribution, 98.5% is concentrated in San Miguel de Cozumel. The localities that occupy a second place in number of inhabitants are El Cedral, San Lorenzo and Huerto Familiar, with around 50 people. There are fifteen locations whose population varies between 7 and 43 inhabitants. Around 500 people live dispersed in 121 locations.

3.2.8.2 Economic activities

Agricultural: Grains, seeds and vegetables are grown in the municipality regions. Most of the products obtained are destined for self-consumption since it is an area defined as temporary and is associated with low yields. The areas destined for agricultural activities are located along Carretera Transversal de Cozumel (Cozumel Transversal Highway) and in El Cedral population center. Some of the harvested vegetables in the region are corn, pumpkins, beans and chili.

Livestock farming: Livestock farming in the area has had a decrease in recent years as well as its dynamics and number of species. The areas destined for these activities are in adjacent areas to the agricultural ones along Carretera Transversal de Cozumel, El Cedral and in some pastures in San Gervancio area.

Fishing: Cozumel Island is the third municipality with the largest fishermen number in the region. Main products include lobsters, snails, and low incidence of shellfish. Fishermen are usually grouped in cooperatives, while in a small proportion, they are usually arranged in distributors.

Tourism: This is Cozumel Island's main economic activity. Land uses suitable for tourism are located in the north and south of the island west coast, as well as in the coastal urban area part.

3.3 Phase 3: identification and assessment of impacts

The environmental impact assessment (EIA) is an environmental policy instrument whose main purpose is designing strategies for the regulation of activities that impact (positive or negatively) on terrestrial and marine environmental systems. These strategies become functional when it is possible to prevent or regulate the changes that the environmental systems' structure, composition and function may suffer due to natural or anthropic causes.

The Leopold matrix methodology was used, which is a tool that allows for visualizing the OC-OTEC plant operation impacts in a practical way. In **Figures 3** and **4**, the Leopold matrix is presented; activities that might have impact (characterized in phase one) are arranged in columns, additionally the environmental inventory, which is represented in rows (characterized in phase two). In some cases, there will be a convergence between rows and columns, which will lead to an evaluation; for this case, three items will be evaluated:

1. The sort of effect, represented with a positive or negative sign (+/-), which indicates if the impact will have benefits (positive) or damages (negative).

2. The impact magnitude, indicated in the upper left, is the spread of the impact or scale impact. The magnitude is scored from 1 to 10 (1 if the alteration is minimal and 10 if it is maximum).

3. The intensity degree or incidence degree, which means how important is the impacting action on a factor from the environmental inventory, is placed in the lower right and is scored from 1 to 10.

			ACTIVITIES THAT MIGHT HAVE IMPACT																			
			1	2	3	4	5	6	7	8	9	10	11	12	13	14	15	16	17	18	19	20
ENVIRONMENTAL INVENTORY	AIR	Wind																				
		Hurricane																				
		Atmospheric contamination			+7/1			-3/2														
		Weather						-2/1														
	SOIL	Type of soil											-1/3								-2/1	-1/1
		Potential land use																				
	GEOLOGY																					
	WATER	Ocean water		-3/1		-3/1	-3/1	-6/2	-8/9	-6/3		-9/8	-3/4				-10/9					
		Surface water																				
		Ground water		+5/3																		
		Geohydrology																				

Figure 3.
Leopold matrix made for the OC-OTEC PLANT project (part 1).

ENVIRONMENTAL INVENTORY		ACTIVITIES THAT MIGHT HAVE IMPACT																			
		1	2	3	4	5	6	7	8	9	10	11	12	13	14	15	16	17	18	19	20
	LANDSCAPE	-3/3																			
	FLORA	-4/2					-2/2						+3/1								
POPU	Demographics	+1/8	+2/8	+6/8	+4/1	+1/1							-1/1	-3/2				-5/5			
POPU	Economic activities	+1/6	+1/6	+1/6	+1/4	+1/4							-1/1								
FAUNA	Sea Fauna							-8/7	-7/7	-4/1	-3/2	-7/7			-2/1	-7/7					
FAUNA	Birds	-2/1																	-2/2		

1	Energy source
2	Freshwater production
3	Production of cooling systems
4	Mineral production
5	Lithium production
6	CO_2 emissions
7	Dragging nutrients to the Surface influences
8	Artificial reef effect / contamination by heavy metal and salts influences
9	Organism drag and compression influences
10	Redistribution of ocean water bodies influences
11	Impact by organic antifouling chemicals influences
12	Noise
13	Lighting
14	Turtle nesting alteration
15	Brine discharge
16	Cause sociocultural impacts
17	Significant public controversy
18	Migration routes interruption
19	Waste
20	Sanitary discharges from the station

Figure 4.
Leopold matrix made for the OC-OTEC PLANT project (part 2) and activities that might have impact explained.

A lack of convergence between the rows and the columns will imply that there is no continuous affectation.

As a result of the Leopold matrix made, we have a total of 43 recognized impacts, explained below:

1. Activity # 1 Energy source interacts with the environmental inventory, causing the following impacts:

    a. The first impact is negative and corresponds to the landscape transformation due to the electric energy transportation (recorded with magnitude 3) toward the population center, which would be through a power line and electricity posts, modifying the landscape in about 2.92 km (distance between the plant and the urban center of Cozumel Island). Intensity degree is 3 because the landscape modification will occur during the construction process. Therefore, reflected intensity during the operation will be relatively low.

    b. During the OC-OTEC plant operation, maintenance of the exteriors is contemplated, so the natural flora will be negatively affected with a 4 magnitude and a 2 intensity degrees due to the modification to the plant species so it is planned not to greatly affect protected species.

    c. The demography will have a positive impact since the inhabitants' projection who can have access to electric energy coming from the OC-OTEC PLANT project in a calendar year amounts to 1772 who corresponds to 2% of the total population (magnitude 1). Although the magnitude is relatively low, it is considered an 8 intensity degrees since the impact it generates on the population's perspective will increase as the character of natural resources use becomes known.

    d. The economic activities will have a positive impact contemplating that the generation of jobs by the production of electric energy through the

OC-OTEC PLANT project is reduced because the workforce is five employees (magnitude 1). The intensity degree is six since the jobs will be filled by personnel in charge of the plant, which will become areas of opportunity for professionals that cover the work profile.

e. Migratory birds whose routes contemplate the polygon will have an impact with magnitude 2 since some species could collide with infrastructure. The intensity degree is 1 because the OC-OTEC PLANT project does not interfere within the migratory routes of the birds that were studied, so it becomes an area of opportunity for future studies.

2. Activity #2 Freshwater production influences the following environmental factors:

   a. In the marine waters, it has an influence of a magnitude 3 because marine water extraction is carried out in the surface and in the depth, which a through time could become a significant impact. Intensity degree is 1; compared to the existing water bodies and ocean distribution, the extraction of seawater that is carried out for the OC-OTEC PLANT operation is not significant being as there is a return of the same.

   b. According to our calculations, an OC-OTEC PLANT generates 1,485,586,365 l of desalinated water per day; therefore, the extraction of fresh water through wells would be reduced, generating a positive impact with magnitude 5 and 3 intensity degrees because the demand for the use of wells on Cozumel Island would be reduced, but not 100%, since the OC-OTEC PLANT could only supply fresh water to 3.24% of the inhabitants per year. On the other hand, not all residents of the island would adapt to the use of new technologies; in addition, there is no exact count of clandestine wells on the island.

   c. The demography will have a positive impact with magnitude 2 since an OC-OTEC PLANT could only satisfy a total of 10,194 people with fresh water, which in comparison to the total population of the island that is 86,415 inhabitants represents only 11.79%. Intensity degree is 8 since it implements the use of innovative alternatives for the production of fresh water, which can serve as an example for the development of projects.

   d. Economic activities have a positive impact with magnitude 1 by implementing the use of innovative alternatives for freshwater production, which can serve as an example for project development. Intensity degree is 6 since the job profiles that are sought to work in an OC-OTEC PLANT are with high academic degrees (master's, doctorate, and specialties).

3. Activity #3 Production of cooling systems influences:

   a. In relation to atmospheric contamination, magnitude 7 is positive since Chlorofluorocarbons (CFCs) that come from conventional cooling systems (air conditioners) are part of the Greenhouse Gases that cause global warming. The production and installation of "natural" cooling systems from the OC-OTEC PLANT would reduce the effect caused by hotel rooms that have such conventional systems on Cozumel Island. Although the production of the cooling system would reduce the amount of CFCs when replacing conventional systems, emissions to the atmosphere that are simply based on water (1 intensity degree) are still being produced.

b. The demography would have a positive impact with magnitude 6 by being able to satisfy a large number of people on the island from cooling systems as around 337.5 tons per year would be generated. The refrigeration system would supply the population, replacing several conventional systems by means of a production coming from an alternative energy (8 intensity degrees).

c. The economic activity would have a positive impact because installation of the cooling system to homes or hotels would generate jobs at the same time as it would be an ecotourism environmental option for those who acquire it (magnitude 1). It is considered that the generation of jobs and capital, by providing an ecotourism environmental option, would come from an alternative energy that would supply lodging centers on the island (6 intensity degrees).

4. Activity #4 Mineral production influences:

    a. Marine waters will have an impact of a negative nature with a magnitude 3 because the extraction of minerals from marine waters could cause an imbalance to marine environmental systems in the future, so it becomes an area of opportunity for future studies. Intensity is considered insignificant when obtaining degree 1 because there is a circulation of oceanic waters in the area so the damage will be minimal; even with this, the study in the future should not be ruled out.

    b. The demography will have a positive impact with magnitude 4. The water that is extracted from the depths is rich in minerals essential for the human body so that nearly 15,000 bottles that would be generated would benefit the population of Cozumel Island. Being a product that is not marketed in the area, it could have a low yield (1 intensity degree).

    c. The economic activity will have a positive impact with magnitude 1 because the staff of collaborators in an OC-OTEC PLANT is very small, so it would only benefit a few people, and the job profiles that are sought to work in an OC-OTEC PLANT are with high academic degrees (4 intensity degrees).

5. Activity # 5 Lithium production influences:

    a. In marine waters, it will have a negative impact with magnitude 3 since the extraction of lithium chloride from marine waters could cause an imbalance to marine environmental systems in the future, so it becomes an area of opportunity for future studies. While the intensity degree is 1 and considered insignificant, because there is a circulation of ocean waters in the area that causes the damage to be minimal, even taking this into consideration should not rule out studies in the future.

6. Activity #6 CO2 emission influences:

a. Atmospheric contamination, since emissions from the OC-OTEC PLANT contribute to the increase of this factor; however, when compared with those resulting from a thermoelectric plant, they are lower, resulting in magnitude 3 and a negative nature. Dispersion can lower this, ensuing to 2 degrees of intensity.

b. Climate, as CO_2 emissions are recorded in the Kyoto Protocol as 1 of the 6 greenhouse gases (GHG) that enliven climate change and could compensate for the occurrence of typical meteorological phenomena in the area, is a negative impact but with a magnitude of 2. Being minimal quantities, which are dispersed, they contribute to an increase of climate change, leading to 1 degree of intensity.

c. Marine waters, because the emissions come both from the extraction of these and from the operation of the OC-OTEC PLANT, which manage to spread on the sea surface (negative impact and magnitude 3). On the other hand, when CO_2 is found in the ocean in a natural way, the minimum emissions of said compound, when diffused in the ocean and the marine surface, would not significantly unbalance the marine environment (2 intensity degrees).

d. Flora, with a magnitude of 2 and negative impact, since only the one closest to the plant would be affected, since being a short distance away, they would be the main captors of the emissions. It receives 2 intensity degrees since despite being minimal emissions to the vegetation closest to the plant, it is considered that environmental stress may occur, as a result of an increase in the amounts of CO_2 normally collected.

7. Activity #7 Dragging nutrients to the surface influences:

a. Marine waters: Although the deepwater extraction zone for the OC-OTEC PLANT is not large, there would be a constant alteration of nutrients to the same ecosystem by the movement of nutrients, which could be dispersed in the surrounding areas thanks to the currents, resulting in a negative impact with magnitude 8. The characteristics of the deepwater zone, from which the necessary water flows will be extracted, are different from the shallowest areas where they would be reinjected, producing a harmful effect by altering the amounts of normal nutrients in both areas for the functioning of the ecosystem (eutrophication), so its intensity degree is 9.

b. The marine fauna will have a negative impact since some marine species live in specific conditions of the area where they are found, and introducing different amounts of nutrients (or increasing them) can cause alterations to these organisms and, therefore, throughout the trophic chain of the own and surrounding ecosystem (magnitude 7). On the other hand, when most of the marine fauna are found at different levels of the area, those that are close to the OC-OTEC PLANT can affect the species belonging to its trophic chain by traveling at different levels (8 intensity degrees).

8. Activity #8 Artificial reef effect/contamination by heavy metal salts influences:

a. Marine waters: With magnitude 6 as negative impact, the protective coatings that are placed in the pipes inlets that prevent the generation of artificial reefs contain components that are in direct contact with the oceans and are made from soluble solid compounds (salts of heavy metal). The most commonly used are Copper (Cu), Mercury (Hg), Zinc (Zn) and Arsenic, which can alter the sustainability of trophic chains causing risks to the environment and society. In the same way, it obtains 3 intensity degrees, since the protective coatings are not placed in all the pipes, only in the inlet nozzles.

b. Marine fauna, negative impact and with magnitude 7: The majority of marine species live in specific conditions of the area where they are; therefore, when introducing heavy metals, an alteration in these organisms can occur and, therefore, in the whole trophic chain of the ecosystem. The purpose of placing heavy metals in the nozzles of the pipes is beneficial to the plant, since the formation of artificial reefs is avoided, and maintenance costs are reduced. On the other hand, it would become a relevant issue if, due to corrosion, there was a leak or spillage of heavy metals into the ocean, resulting in 7 intensity degrees.

9. Activity #9 Organism drag and compression influences:

a. Marine fauna: Because the marine species that inhabit the zones of extraction of marine waters are latent subjects to enter the pipes and perish, that is a reason why it is considered a negative impact and magnitude 4. Also, there are technologies that prevent the passage of marine species, such as gratings or retention ponds, where they can be manually removed, resulting in 1 intensity degree.

10. Activity #10 Redistribution of ocean water bodies influences:

a. Marine waters: The water rejection discharges that are used as working fluid within the operation of the OC-OTEC PLANT project are at a temperature in the condenser of 10.5°C and in the evaporator of 21.5°C, which is different from the one that occurs in the ocean at the time of the discharge, which causes to score with this impact as negative and a magnitude of 9. Similarly, at the time of the discharge of water rejection, it will come into contact with the ocean, producing a shock term that will dilute until reaching a balance within the water column, thus resulting in 8 intensity degrees.

b. Marine fauna, negative impact and magnitude 3, since the temperature at 150 m is 24.5°C, while the discharge of water from the evaporator is 21.5°C, that small temperature change of 3° C can favor the relocation of fish in search of an optimal ecosystem for their reproduction, feeding and growth. Likewise, the OC-OTEC PLANT is located in the western part of Cozumel Island, so that the said area does not converge with coral species that may be affected (2 intensity degrees).

11. Activity #11 Impact by organic antifouling chemicals influences:

a. The type of soil, as a negative impact and with magnitude 1, during the OC-OTEC PLANT maintenance, only antifouling chemicals will be used in the flash evaporator; therefore, only in case of any leakage, spillage or mishandling of products could this affect the earth's crust. The only chemical used in the OC-OTEC PLANT is chlorine and its greatest impact would be only if there are critical leak situations. Chlorine reacts docilely with the chemical components of the soil in the region forming chlorides that, depending on their solubility, are easily washed with water. A spill of liquid chlorine could temporarily freeze the affected soil area (3 intensity degrees).

b. Marine waters, considered as negative impact and with magnitude 3, would only be affected if at the time of maintenance (washing) of the evaporator, this remains with chlorine residues that are dragged into the ocean. It is not an impact that happens frequently only in case there is a mishandling of chemicals so the way it would affect would be the following: chlorine rapidly hydrolyses in water forming hypochlorous acid and hydrochloric acid. Free chlorine reacts quickly with organic compounds. Concentrations of up to 0.05–0.15 mg/l cause significant changes in the composition of marine phytoplankton species, thus obtaining 4 intensity degrees.

c. Marine fauna, as a negative impact with magnitude 7, since chlorine is a substance considered toxic, therefore, it affects marine species, and thus produces effects on marine fauna, which would affect the food chain and the ecosystem in general, giving it 7 intensity degrees.

12. Activity #12 Noise influences:

a. Flora, as a positive impact and magnitude 3: The operation of the OC-OTEC PLANT project has noise emissions that are below those established in Mexican standards, to this we must add that being surrounded by plant species creates a natural barrier that prevents the propagation of noise to neighbors. Similarly, the intensity is negligible because the OC-OTEC PLANT alone does not exceed Límites Máximos Permisibles (Maximum Permissible Limits) stipulated in NOM-080-STPS-1993 (1 intensity degree).

b. Demography, negative impact and only with magnitude 1, since there are no urban centers of population nearby. There are shops 197.48 meters to the northeast and 183.18 meters to the northwest, but because they are separated by a large distance where natural dissipation barriers are found, the OC-OTEC PLANT project will not affect neighboring buildings. On the other hand, flora is a natural barrier that will prevent the transport of noise and vibrations, so it will not affect neighboring buildings, thus obtaining 1 intensity degree.

c. Demography, of a negative nature and only with magnitude 1, because there are no urban centers of population nearby. There are shops 197.48 meters to the northeast and 183.18 meters to the northwest, but because they are separated by a large distance where natural dissipation barriers are found, the OC-OTEC PLANT project will not affect neighboring buildings. On the other hand, flora is a natural barrier that will prevent the transport of noise and vibrations so it will not affect neighboring buildings, thus obtaining 1 intensity degree.

13. Activity #13 Lighting influences:

a. Demography, being a negative impact with magnitude 3, because it could be the cause of causing disturbances at two different levels: Internally, it could affect work comfort and externally in vehicular traffic. Norma Oficial Mexicana NOM-025-STPS-2008, Condiciones de iluminación en los centros de trabajo (Official Mexican Standard NOM-025-STPS-2008, lighting conditions in the workplace), establishes only Límites Mínimos (Minimum Limits) for the correct performance in the workplace as well as a section in which the employee must be informed about the risks of glare, so it is ambiguous how to standardize it. While on the outside, there are no regulations in force that stipulate Límites Máximos Permisibles for outdoors. However, the intensity degree is 2 because although it could generate disturbances outdoors as well as indoors they would be in very isolated cases.

14. Activity #14 Turtle nesting alteration influences:

a. Marine fauna: It is considered as a negative impact and with magnitude 2 since the OC-OTEC PLANT has artificial lighting at night time due to safety reasons, but due to its location, it does not alter the nesting of turtles. Also, the OC-OTEC PLANT is located in a site that is not considered to be the site of nesting turtles of any of the 6 species, so it only has 1 intensity degree.

15. Activity # 15 Brine discharge influences:

a. Marine waters, with a negative nature and magnitude 10, due to the excess of sodium chloride in the brine resulting from the operation of the OC-OTEC PLANT, which when discharged to the marine waters, will increase the normal concentrations of salts in the ocean. Its intensity degree is 9 due to the different sea currents that interact in nearby marine areas and high salt concentrations can cause great dispersion and affectation to different ecosystems.

b. Marine fauna has a negative impact and magnitude 6. The increase in salinity of certain waters disturbs the osmotic balance of marine species with their environment, which in turn produces dehydration of the cells and decrease in the pressure of the turgidity and can lead to the death of the species. Marine species close to the OC-OTEC PLANT are the most likely to be affected by the increase in salt concentrations, which can lead to the said alteration in these organisms due to their participation in the food chain and when they reach other places. This results in 6 intensity degrees.

16. Activity # 16 causing socio-cultural impacts does not interfere in any way with any type of activity described.

17. Activity # 17 Causing significant public controversy influences:

a. Demography with a negative nature and magnitude 5. This is due to the fact that there are an infinite number of aspects (lack of market studies, lack of job opportunities for local inhabitants, irregularities in construction, damage to ethnic groups and the public image) due to that the society of Cozumel Island can take arguments to oppose the operation of the project. It depends largely on the information provided to the inhabitants of the island about the OC-OTEC PLANT project, resulting in 5 intensity degrees.

18. Activity # 18 Disruption of interruption of migration routes influences:

    a. Birds, as a negative impact and magnitude 2, since the exterior lighting of the OC-OTEC PLANT project will be lit at night due to internal security issues, which could generate disturbances to migratory species, even with this some bird species were characterized and it was verified that none of it has a migration route close to the facilities of the OC-OTEC PLANT project. This is not an impact that happens frequently; however, it must be taken into account for future projects, thus having 2 intensity degrees.

19. Activity # 19 Waste influences:

    a. Type of soil, as a negative impact and magnitude 2. The OC-OTEC PLANT project is a small generator of Urban Solid Waste (MSW) since an annual projection would produce an approximate 6 kg/day or 2190 kg per year. Intensity degree is low (1) because an external company will be responsible for the collection and final disposal of it, so the only way for contamination to exist is due to its poor handling.

20. Activity # 20 Sanitary discharges from the station influences:

    a. Type of soil, as a negative impact and only magnitude 1, since the wastewater in the OC-OTEC PLANT project will come from a bath that will be connected to the sewer system of Cozumel Island. Its intensity degree is 1 because the workforce is made up of 5 employees, so the wastewater will be minimal.

3.4 Phase 4: proposal for corrective and preventive measures

According to the Leopold matrix corresponding to the OC-OTEC PLANT operation project, it was found that of the 20 impacts identified as likely to occur during the operation of the project, only 4 of them are significant because they have a greater amount of negative impacts and have a greater weighting over other impacts, in order to reduce the probability of causing greater damage to the environment and components of the environmental inventory, a series of corrective and/or preventive measures that can be used in the facilities of the OC-OTEC PLANT project.

3.4.1 Dragging nutrients to the surface

A measure for the reduction of impacts, caused by pumping cold water from the depths of the ocean carrying nutrients, could be the alternative uses for wastewater (water with nutrients), to avoid reinjection into the ocean. One of the options would be to filter to separate the nutrients from the water, which would consist of passing the liquid through a filter to retain some of its components; in this way, we would have two waste materials (water/nutrients) that could be used separately for other purposes. For example, water through a purification process can be distributed and sold on the island. On the other hand, nutrients can be dehydrated in the sun and sold dry as salt or fertilizer; in this case, a relevant study would have to be done.

3.4.2 Redistribution of ocean water bodies

Alternative uses can be given to rejection waters from the evaporator and condenser in order to reduce the impact on the environment. Some of the options are as follows:

1. Temperature regulator: The cold rejection water will pass through a continuous reactor in which temperature regulators will be placed to match the temperature of the water with that of the ocean.

2. Aquaculture: Use of deep sea cold water to grow commercial organism crops such as fish, molluscs, crustaceans and algae that reproduce in cold water in tropical areas.

3. Agriculture (cold water): Cold water from the depths is piped and moves under arable soil to cool it (indirectly). This process helps the plant roots to stay cold and generates condensation of the moisture present near the soil (approximately 10° C). As the water continues to cool, it penetrates through the soil until it reaches the coldest point of the root zone. Plants from temperate and semitemperate areas could be grown in tropical areas, such as the following plant species: strawberries, grapes, carrots, onions, apples and peaches.

3.4.3 Impact by organic antifouling chemicals

The chosen organic antifouling chemical for the OTEC plants maintenance is chlorine, which, together with other chemical compounds (metal ions, ammonia, detergents, polychlorinated biphenyls, etc.), is the main toxin that causes damage to marine ecosystems.

Since this may occur in specific and / or isolated cases due to improper use of the product or by residues contained in the OC-OTEC PLANT evaporator, it is relevant to mention that the establishment of a preventive measure rather than corrective measures is preferable to reduce the presence of pollutants.

The prevention would be carried out by the development, implementation, use and/or updating of an "Environmental Management Plan," which will help to avoid, mitigate and/or progressively compensate the acting of personnel in extraordinary situations. Some of the activities described in the environmental management plan would be:

1. In compliance to Norma Oficial Mexicana NOM-005-STPS-1998, Relativa a las condiciones de seguridad e higiene en los centros de trabajo para el manejo, transporte y almacenamiento de sustancias químicas peligrosas (Official Mexican Standard NOM-005-STPS-1998, concerning the conditions of safety and hygiene in workplaces for the handling, transport and storage of hazardous chemical substances), which stipulates conditions that will help to prevent and protect the health of workers, avoid damage to work facilities and cause adverse damage to the environment.

2. Implementation of chlorine neutralization systems (or as they are also known as scrubbers) that allow emergency emergencies to be leaked, by passing the

chlorine gas through an extractor (Venturi scrubber), which contains a dry or moisturizing neutralizing medium that will help dissipate said product in nontoxic amounts to the environment.

3. Emergency plan to react on time if that event occurs with training for employees is contemplated.

3.4.4 Brine discharge

As alternatives to avoid reinjection into marine waters, and recognizing the importance of mangroves throughout the state area, a preventive measure of the impacts of brine discharge is their collection for use in mangrove reforestation processes. These processes can be focused on direct reforestation in disturbed mangrove areas or reforestation with nursery seedlings, of which, in the latter, it is important to preadapt them to conditions similar to the place where they are sown to ensure greater survival, with salinity being especially important, so it is in this case where the use of the necessary amounts of the "natural" brine from the OC-OTEC PLANT operation would come in. Several studies indicate that *A. germinans*, or black mangrove, has a high tolerance to salinity, so it is highly recommended for use from seedlings of that species. Following the same line of alternatives, the brine can be transferred to be used in salt-obtaining equipment or for forced osmosis systems (osmosis by means of difference in salt concentrations), taking into account that the final residues of these processes also can be used for mangrove reforestation.

On the other hand, as some methods to minimize the impacts of the discharge, as no alternatives to reinjection are contemplated, various desalination projects recommend predilution processes with seawater or discharge through diffusers, to equalize marine concentrations.

4. Discussion

Among the factors to consider for OTEC commercialization, the environmental impacts exists as well as public acceptance, so it is necessary to address potential impacts and risks of OTEC during a plant's planning, design, construction, commissioning and operation, from a technical and scientific point of view, in order to explicitly present the impacts, potential risks and mitigation strategies to the interested parties and general population, all the aforementioned being, the initial starting point of this document, focused on the open-cycle OTEC plant operation [33].

One way to identify and evaluate environmental impacts is through instruments such as checklists or various types of matrices (Leopold, Cause and Effect, Batelle-Columbus, etc.), which are applicable to the different phases of the realization of a project (construction, operation and abandonment). For this project, the Leopold matrix was used, which has evaluated projects such as a thermoelectric plant [34] in which it has been very useful for measuring environmental impacts.

Globally, environmental impact assessments, or similar, focused on OTEC technology, have been developed with various tools and have concluded in different impacts, as well as mitigation and prevention strategies.

For example, a Final Environmental Impact Statement prior to 1981 is available in Hawaii [35], which must be updated to current oceanographic and engineering standards. Based on the high research done on Hawaii's oceanography, a report is presented that highlights some of the most important contributions in terms of OTEC development, as well as the gaps in knowledge, and finally proposes a protocol for monitoring the environmental baseline, focusing on a set of ten chemical oceanographic parameters relevant to OTEC and addresses the gaps in the knowledge of the ecology and oceanography of the area chosen for the development of OTEC.

In Beijing, China, it is considered that site selection is a key factor for the success of an OTEC project, and in turn, it is a problem of multiple-criteria decision-making (MCDM) with a series of conflicting criteria involved. Classic MCDM methods tend to fail for these projects [36]. Therefore, an extended TDCM method is proposed and a new decision-making framework is established to select the optimal location. First, a complete system of evaluation criteria is established for the OTEC site's selection from sustainability point of view with 16 criteria involved, among which it mentions that environmental sustainability evaluations are necessary to assess the ecological and environmental impact of carbon in the environment before decision-making, resulting in the criterion named possible environmental effect (C51), which covers biological and environmental impacts during the operation of the plants; however, their method does not offer any kind of recommendation or mitigation plan for such impacts.

Another case is from the island of Martinique, in France, where bibliography was studied in detail, both for the evaluation of the site and for the impacts of other projects in the world with similarities relevant to OTEC; a thematic synthesis was subsequently carried out for each physical, biological and human theme that could be affected by the project, resulting in a total of 28 issues, all of the above with the objective of establishing priorities for specific evaluations [37]. This last case is shown as the most similar to this document, since it is known, it involved a documentary investigation on the physical and biological environment of the study site in Cozumel and its impacts, throwing 37 topics addressed (20 activities that might have impacts and 17 environmental inventory factors) that are not so far from the number of issues in the case in question and taking into account that in both, priorities were selected for further study.

On the other hand, of the different impacts described for the OC-OTEC project, it is possible to contrast some impacts from data obtained with those of other similar works, such as the production of electrical energy knowing that of the 86,415 inhabitants of the island from Cozumel, about 1772 would benefit. The distribution of energy through an OTEC PLANT is a viable option in Martinique, France, as it would benefit 2187 inhabitants, a number quite close to the case of Cozumel Island [38]. Likewise, there are other works [39] where we talk about the environmental impacts corresponding to an open-cycle OTEC plant, such as atmospheric effects from CO2 and marine impacts, which are also discussed in this work.

As mentioned previously [33], one of the factors to consider for the commercialization of OTEC is the degree of public acceptance, which, by reviewing different impact evaluations, can be found to be of no significant importance. For example, in 2012, Secretaría de Medio Ambiente y Recursos Naturales (SEMARNAT, acronym in Spanish for Ministry of Environment and Natural Resources) rejected a project for the implementation of a Wind Farm in Cozumel due to citizen opposition, from businessmen and academics, since they argued that it put the habitat, biodiversity and species at risk in serious danger of extinction. In this civil power lies the importance of knowing the social environment at the time of carrying out projects likely to affect the environment.

Around the world, research on OTEC plant technology has been developed in places like Mexico (Bárcenas, 2014), Hawaii (Vega, 1991), and India (Ravindran &

Raju, 2002), among others. Finally, the places with plants currently in operation are Kumejima in Okinawa (Japan); Saga University (Japan); Goseong (South Korea); Reunion Island (France); and Makai, Island of Hawaii (United States) [40].

5. Conclusion

The elaborated Leopold matrix results made it possible to establish a series of preventive and corrective measures aimed at protecting the environment of the polygon studied. Despite having gathered enough information for the EIA on both the matrix and the environment of Cozumel Island, as well as the impacts produced by the operation of the plant, the low existence of recent data represented a limitation in that aspect. We had to work mainly with data from management programs and preliminary evaluations of the different protected natural areas of the island, being quite useful in aspects of protection, as well as for the establishment of an OC-OTEC plant. The importance of developing these studies on environmental impact is highlighted, since considering their findings, such as the protection of mangrove areas, the existence of land uses determined by law to avoid this kind of construction in specific zones, a 51% approval of a population sample, and sharing the benefits from water desalination, the environment and the immediate society are contemplated, thus leading to renewable energies such as OTEC to an advanced sustainability.

Acknowledgements

We would like to express our gratitude to Centro Mexicano de Innovación en Energía del Océano (CEMIE-O), especially its participants MSc. Luis Mariano Sánchez Avelar and MSc. Alejandro García Huante, for their interest in this research and advices provided for its development. Likewise, we appreciate the valuable collaboration of Departamento de Inglés (Language Department) and Departamento de Desarrollo Humano (Humanities Department) belonging to Universidad del Caribe, especially to Ms. Jane Acosta Holmes, the Language Department head, and MPhil Roberto Parra Dorantes, research professor at the Humanities Department, as well to Elvira Acuña Gonzalez, Thelma Canto Burgos and Magdalena Avila Pardo, professors in the Language Department, for their counseling, support and commitment for the translation of this chapter.

Author details

Enrique Celestino Carrera Chan, María Fernanda Sabido Tun,
Juan Francisco Bárcenas Graniel* and Estela Cerezo Acevedo
Universidad del Caribe, Centro Mexicano de Innovación en Energía del Océano (CEMIE-O), Cancún, México

*Address all correspondence to: jbarcenas@ucaribe.edu.mx

References

[1] Ocean Energy Systems - International Energy Agency. 2017. 2017 Annual Report: An overview of Ocean Energy Activities in 2017. Available from: https://report2017.ocean-energy-systems.org/

[2] Bárcenas J. Evaluación del Potencial energético Renovable del Caribe Mexicano. Master's Degree Thesis Not Published. México, DF: Instituto de Ciencias del Mar y Limnología. UNAM; 2014

[3] Gómez D. Evaluación de Impacto Ambiental. 2nd ed. Madrid, España: Editorial agrícola española; 1998

[4] Ravindran, M. & Raju, A. The Indian 1 MW demonstration OTEC plant and the development activities. Oceans '02 MTS/IEEE. 2002. DOI: 10.1109/oceans.2002.1191877

[5] Aviña H. Factibilidad de Instalar Una Planta OTEC (Ocean Thermal Energy Conversion) en Las Costas de México (Bachelor's Thesis). México, DF: Universidad Nacional Autónoma de México; 2007

[6] Institut français de recherche pour l'exploitation de la mer (IFREMER). 2014. APEX Profiling Float SBE_6587. (Database). Available from: http://www.ifremer.fr/co-argoFloats/float?detail=false&active=false&ocean=P&lang=en&ptfCode=4901644

[7] Vega L. OTEC overview, by LA Vega, Ph.D. Ocean Thermal Energy Conversion (OTEC). OTEC News. 1999. Available from: http://www.otecnews.org/portal/otec-articles/ocean-thermal-energy-conversion-otec-by-la-vega-ph-d/#open

[8] International Bathymetric Chart of the Caribbean Sea and the Gulf of Mexico (IBCCA). 2016. Conjunto de datos vectoriales Carta batimétrica internacional del Mar Caribe y Golfo de México (IBCCA) Hoja 1-07. (Base de datos). Available from: https://datos.gob.mx/busca/dataset/conjunto-de-datos-vectoriales-carta-batimetrica-internacional-del-mar-caribe-y-golfo-de-mexico-/resource/fdafd79a-326f-4f52-b5c4-8329ce1251b2?inner_span=True

[9] Comisión Nacional de Áreas Naturales Protegidas (CONANP). 2017. Reservas de la Biosfera, Parques Nacionales, Monumentos Naturales, Áreas de Protección de Recursos Naturales, Áreas de Protección de Flora y Fauna, Santuarios de CONANP (Database). Available from: https://datos.gob.mx/busca/dataset/informacion-espacial-de-las-areas-naturales-protegidas-181

[10] Avery W, Wu C. Renewable Energy from the Ocean. A Guide to OTEC. New York: Oxford University Press; 1994. pp. 446

[11] Vega L. Economics of Ocean Thermal Energy Conversion (OTEC): An Update. Offshore Infrastructure Associates, Inc. [thesis]. USA, Hawaii: University of Hawaii; 2010

[12] Secretaria de Energía. Evaluación rápida del uso de la energía Cozumel, Quintana Roo, México. [Internet]. 2016. Available from: https://www.gob.mx/cms/uploads/attachment/file/170882/7__Cozumel.pdf [Accessed: 10 March 2019]

[13] Secretaria de Energía. Programa de desarrollo del sistema eléctrico nacional 2015-2029. [Internet]. 2015. Available from: https://www.gob.mx/cms/uploads/attachment/file/54139/PRODESEN_FINAL_INTEGRADO_04_agosto_Indice_OK.pdf [Accessed: 17 February 2019]

[14] Instituto Nacional de Estadística y Geografía. Anuario estadístico y

geográfico de Quintana Roo 2013. [Internet]. 2013. Available from: http://bibliotecasibe.ecosur.mx/sibe/book/000011171 [Accessed: 05 January 2019]

[15] Secretaria de Energía (SENER). Programa de desarrollo del sistema eléctrico nacional 2018-2032. México: Secretaria de Energía; 2018 pp. 107, 110, 175, 244, 284

[16] Instituto Nacional de Estadística y Geografía (INEGI). Anuario estadístico y Geográfico de Quintana Roo 2016. Quintana Roo, México: Autor; 2016

[17] Instituto Nacional de Estadística y Geografía (INEGI). Viviendas particulares habitadas por municipio y su distribución porcentual según disponibilidad de energía eléctrica Al 15 de marzo de 2015. Dirección General de Estadísticas Sociodemográficas. (Encuesta intercensal); 2015. Available from: www.inegi.org.mx

[18] Instituto Nacional de Ecología (INE). 1998. Programa de Manejo Parque Marino Nacional Arrecifes de Cozumel. Available from: https://www.conanp.gob.mx/que_hacemos/pdf/programas_manejo/cozumel.pdf

[19] Instituto Nacional de Estadistica y Geografia. Anuario estadistico y geografico de Quintana Roo 2016. [Internet]. 2017. Available from: https://www.datatur.sectur.gob.mx/ITxEF_Docs/QROO_ANUARIO_PDF.pdf [Accessed: 13 January 2019]

[20] Alcérreca I. Plan de Acción Climática Municipal (PACMUN) (Power Point dispositive). 2012. Available from: http://www.pincc.unam.mx/congresonacional2012/sis_admin_pres/archivos_2012/pacmun_octubre_2012.pdf

[21] García E, Comisión Nacional para el conocimiento y uso de la Biodiversidad (CONABIO). 2004. Climas. (Base de datos) Available from: http://www.conabio.gob.mx/informacion/metadata/gis/clima1mgw.xml?_xsl=/db/metadata/xsl/fgdc_html.xsl&_indent=no

[22] Orellana R, Nava F, Espadas C. El clima de Cozumel y la Riviera Maya. Biodiversidad acuática de la isla de Cozumel. 1st ed. México: Centro de Investigación Científica de Yucatán; 2007. p. 422

[23] Spellman F. The Science of Renewable Energy. 2nd ed. Florida: CRC Press; 2016

[24] Secretaría de Medio Ambiente y Recursos Naturales. 1994. Norma Oficial mexicana NOM-081-SEMARNAT-1994 Que Establece los Límites Máximos Permisibles de Emisión de Ruido de las Fuentes Fijas y su Método de Medición. Available from: http://biblioteca.semarnat.gob.mx/janium/Documentos/Ciga/agenda/PPD02/081.pdf (Norma Oficial Mexicana publicada originalmente en el Diario Oficial de la Federación el 23 de abril de 2003)

[25] Instituto Nacional de investigaciones Forestales y Agropecuarias (INIFAP)–Comisión Nacional para el Conocimiento y Uso de la Biodiversidad (CONABIO). 1995. Edafología. (Database). Available from: http://www.conabio.gob.mx/informacion/metadata/gis/eda251mgw.xml?_httpcache=yes&_xsl=/db/metadata/xsl/fgdc_html.xsl&_indent=no

[26] Bautista F, Maldonado D, Zinck AJ. La clasificación maya de suelos. Ciencia y desarrollo. 2012;**260**:64-70

[27] Butterlin J, Bonet F. Información básica para la interpretación geohidrológica de la Península de Yucatán. México, DF: Secretaria de Recursos Hidráulicos; 1960

[28] Instituto Nacional de Ecología (INE). 1998. Programa de Manejo Parque Marino Nacional Arrecifes de

Cozumel. Available from: https://www.conanp.gob.mx/que_hacemos/pdf/programas_manejo/cozumel.pdf

[29] Skinner B, Murck B. 2011. The blue planet: an introduction to earth system science (No. 504.3/.7 SKI)

[30] Comisión Nacional de Áreas Naturales Protegidas (CONANP) and Secretaría de Medio Ambiente y Recursos Naturales (SEMARNAT). 2018. Programa de Manejo Reserva de la Biosfera Caribe Mexicano. Available from: https://www.conanp.gob.mx/programademanejo/RBCMPM.pdf

[31] Cámara de Diputados del H. Congreso de la Unión. (21 de mayo de 2013). Acuerdo por el que se dan a conocer los estudios técnicos de aguas nacionales subterráneas del acuífero isla de Cozumel, clave 2305, Estado de Quintana Roo. Available from: http://www.dof.gob.mx/nota_detalle.php?codigo=5299571&fecha=21/05/2013 (Acuerdo publicado originalmente el 21 de mayo de 2013 en el Diario Oficial de la Federación)

[32] Rivera M. Estudio Geofísico e hidrogeológico para un banco de materiales, en Playa del Carmen Quintana Roo, México. [Thesis]. México, D.F.: Universidad Nacional Autónoma de México; 2013

[33] Romero A. Evaluación del potencial maremotérmico en costa Norte y Sur del Estado de Veracruz. (Master's degree thesis). Veracruz, México: Universidad Veracruzana; 2013

[34] Garza, R. 36 CC Baja California III (Sitio: La Jovita), Modalidad: Productor Externo de Energía (PEE). Manifestación de Impacto Ambiental. 2010. Available from: sinat.semarnat.gob.mx/dgiraDocs/documentos/bc/eUstudios/2010/02BC2010E0007.pdf

[35] Comfort C, Vega L. Environmental Assessment for Ocean Thermal Energy Conversion in Hawaii. Available data and a protocol for baseline monitoring. OCEANS'11 MTS/IEEE KONA. 2011

[36] Zhang J, Xu C, Song Z, Huang Y, Wu Y. Decision framework for ocean thermal energy plant site selection from a sustainability perspective: The case of China. Journal of Cleaner Production. 2019;**225**:771-784. DOI: 10.1016/j.jclepro.2019.04.032

[37] Auvray C, Ledoux S, Diaz B, Yvon C, Pouget-Cuvelier A. Environmental impact assessment for an OTEC Plant in Martinique Island. La Houille Blanche. 2015;**2**:60-66

[38] Devault D, Péné-Annette A. Analysis of the environmental issues concerning the deployment of an OTEC power plant in Martinique. Environmental Science and Pollution Research. 2017;**24**(33):25582-25601

[39] Quinby-Hunt M, Wilde P, Dengler A. Potential environmental impacts of open-cycle thermal energy. Environmental Impact Assessment Review. 1986;**6**:77-93

[40] Garduño E, García A, Rodríguez Y, Bárcenas J, Alatorre M, Cerezo E, et al. Conversión de Energía Térmica Oceánica (OTEC) Estado del Arte. 2017. Available from: https://cemieoceano.mx/downloads/libros/CEMIEOceano_Gradiente_Termico.pdf